Docker for Developers

2nd Edition

by Chris Tankersley

php[**architect**] edition

Docker for Developers

php[architect] edition published: August 2017

ISBN - Print:	978-1-940111-56-8
ISBN - PDF:	978-1-940111-57-5
ISBN - ePub:	978-1-940111-58-2
ISBN - Mobi:	978-1-940111-59-9
ISBN - Safari:	978-1-940111-60-5

Produced & Printed in the United States

Disclaimer

Written by
Chris Tankersley

Layout and Design
Kevin Bruce

Published by
musketeers.me, LLC.
201 Adams Ave.
Alexandria, VA 22301 USA

240-348-5PHP (240-348-5747)
info@phparch.com
www.phparch.com

Table of Contents

To my wonderful wife Giana, who puts up with all the stuff I do, and my two boys who put up with all of my travel.

Thank you to the people that read this book early and provided valuable feedback, including Ed Finkler, Beau Simensen, Gary Hockin, and M. Craig Amiano.

I'd also like to thank the entire PHP community, for without them I wouldn't be where I am today as a developer.

Preface

This book, for me, is a long time coming. I've been following containerization for a long time, and got very interested in it in 2013 after seeing a presentation on dotCloud at the first MidwestPHP conference. I'd had previously used things like chroots and jails in my career, but containers took most of that a step further.

It wasn't too long after that that the Docker project was announced, in fact it was only about a week later, and was started by the very same company I had just seen a presentation on. I started playing with it right away. It was primitive, but it made building Linux containers much, much easier than working in straight LXC.

Over the last few years there has been an explosion of software and Sofware-as-a-Services pop out of the woodwork to deal with Docker, and Docker finally announced what I thought were three major tools for working with Docker - being able to provision boxes easily, being able to cluster them, and being able to easily bring up mult-container environments. Yes, there were tools that did this made by third parties, but in this technological age I did not want to be bound to a particular vendor, and I did not want my readers to be bound by that as well. I wanted everything native.

Today we have almost everything a user needs in Docker. I feel confident that I do not need to burden my readers with a specific vendor and that you, the reader of this book, can get Docker up and running for what you need.

I've written plenty of articles and taught plenty of classes, but I'm proud for this to be my first book. I hope you enjoy it.

Assumptions

I know, I know. Making assumptions makes… well, you know the rest. In any event, I'm writing this book for a developer who is familiar with PHP and is looking at Docker to either help with deployment issues, or development issues introduced through manually managing servers or keeping development environments in line. Maybe you are not happy with using a full virtualization stack like Vagrant, or are having issues with maintaining virtual machines across a wide variety of developers. This book will help you learn Docker, and possibly how it can help you deploy applications and make a better development environment for everyone.

This book will not teach you PHP. In fact, this book really doesn't care what kind of application you are building. I am only using PHP as an example since it is a well known and understood programming langauge that very heavily leans toward a multi-tiered application. Many modern PHP applications have a web tier, such as nginx, an application tier (PHP), and a database tier like MySQL. These types of applications are well suited for what we will be doing.

This book should work for any type of application that you want to set up in Docker, be it in Ruby, Python, or Go. This book focuses on full stack applications but will easily work with contained applications, such as C programs that are meant to be run from the command line. The ideas are all the same.

At the very least, read through this book if you want to know how to use Docker.

Style Conventions

Throughout the book there will be many times where a command is given as an example. Each command will have a $ preceeding the command, the command itself, and then possibly a result of the command output. These sections will be in a monospace font, and look similiar to the following:

```
$ docker -v
Docker version 17.03.0-ce, build 60ccb22
```

docker -v is the command to run, so if you copy and paste the example into your own terminal make sure to leave off the preceding $. The second line is the output of the command. Some commands may have more lines of output, some none at all or have been ommited for brevity.

Many commands will be too long to properly fit on a single line on the book page and will be broken up using the \ character. You can still copy and paste these commands into most terminals as they will properly process the \ character. If you are manually typing these commands, feel free to leave out the \ character and put the entire command on one like. For example:

```
$ docker run \
    --rm -ti ubuntu \
    bash
```

is equivalent to running:

```
$ docker run --rm -ti ubuntu bash
```

Sample Code

This book has a corresponding set of sample code to make reader's lives easier. To use these code examples, you will need to have either the git[1]

program installed on your computer, or use a tool like Github Desktop[2],

Tower[3],

or an IDE that supports git like Jet Brains PhpStorm[4].

You can download the sample code from Github at *https://github.com/learningcontainers/dockerfordevs*. By using a git repository, I can make it much easier for you to download updates as the book itself updates, and keep all of the sample code in one place. Each sample is stored in a branch, or a separate version, of the repository.

For example, one of the chapters has sample code for creating a Docker Compose file, and will instruct you to switch to the `containerize/compose-file` branch. You can do that by going into the folder with the sample code and running:

```
$ git checkout containerize/compose-file
```

git will switch the code over to the new branch and you will have access to the sample code. For more information on git, check out the main website at *https://git-scm.com* or the documentation for the software you are using.

[1]   git: *https://git-scm.com*
[2]   Github Desktop: *https://desktop.github.com/*
[3]   Tower: *https://www.git-tower.com/windows/*
[4]   Jet Brains PhpStorm: *https://www.jetbrains.com/phpstorm/*

Chapter

1

Containers

The development world goes through many different changes, just like technology in general. New ideas and concepts are brought out from various different places and we integrate them into our workflow. Some of these things are flashes in the pan, and others help revolutionize how we work and make our development lives easier.

About eight years ago virtualization started taking hold in the server room, and that eventually led to the need to run virtual machines locally. It was expensive and slow, and eventually technology caught up. Once virtual machines became disposable, we needed a way to maintain them, and we ended up with Vagrant.

Vagrant, like many tools developers use, is a wrapper around an existing technology. Vagrant makes it easier to download, create, maintain, and destroy virtual machines, but it's not virtualization itself. That's handled by mature programs like Virtualbox or VMWare Workstation/ Fusion. Vagrant puts a single interface on those technologies and allows us to share environments.

This is all well and good, but virtualization takes a heavy toll. Whenever we boot up a virtual machine, we are turning on a second computer (or third or fourth) inside our existing computer.

Each virtual machine needs enough RAM and CPU to run that full little machine, and it takes it from your host machine. See Figure 1-1.

Eight years ago when I started using virtualization day-to-day this was an issue as my desktop machines rarely had more than four gigabytes of RAM and dual cores. Now I'm running a quad core box with twenty gigabytes of RAM and never notice the multiple machines inside of it. The cost of running virtual machines is still high but our machines are large enough to handle them without problems.

As time goes on we are not seeing the huge leaps and bounds in computer CPUs and memory capacity that we have historically seen. With this in mind, developers and system administrators have started to look at containers as way to better utilize what power machines have. Containers are an alternative to virtual machines because instead of booting an entire machine inside of another one, containers section off processes to make them think they are in their own little world.

There are many different ways that containers can be built, but the idea is pretty much the same. We want to seperate processes from each other so that they don't interact and make it easier to deploy systems.

A Basic Container

As I said, this is nothing new. Containers have been around as an implementation detail in Unix-type systems since 1982 with the introduction of `chroot`. What `chroot` does is change the root directory for a process so that it does not see anything outside of the new root directory. Let's look at a sample directory structure:

Figure 1-1

```
 1. /
 2. |-> bin/
 3. |    |-> which
 4. |    |-> bash
 5. |    '-> grep
 6. |-> home/
 7. |    |-> bob/
 8. |    |    |-> bin/
 9. |    |    |    '-> bash
10. |    |    '-> public_html/
11. |    '-> alice/
12. |-> usr/
13. '-> var/
```

Let us say that we have two users, Bob and Alice. Both are regular system users and can SSH into the computers. Alice is a normal user, so when she does cd /, she sees the folders bin/, home/, usr/, and var/. Depending on her privileges she can go into those folders, and call the programs `which`, `bash`, and `grep`. In fact, her shell is `/bin/bash`, so when she logs on that gets called. To her, she has full access to the system, and if you are using Linux, OSX, or any other system that is how it works for you.

Bob, however, is inside of a chroot. We've set his chroot to `/home/bob/` because he is a client of Alice's, and doesn't need full access to the machine. For Bob, when he SSH's in and runs `cd /`, he only sees `bin/` and `public_html/`. He can not see any higher in the directory tree than `/home/bob/` because the system has changed his root from `/` to `/home/bob/`. We've moved Bob off into his own little corner of the world.

This presents a few problems. If his shell is `/bin/bash`, we need to move a copy of that program into his world. So we now have two copies of `bash`, one in `/bin/` and another in `/home/bob/bin/`. We have to do this because Bob can't see anything higher than his chroot, and the regular system `bash` is outside of Bob's root.

We've put Bob inside his own container. He's free to do anything inside his container he wants without impacting the rest of the machine (barring running processes that use all of the system resources like I/O, RAM, or CPU usage). If he screws up and deletes everything by running `rm -rf /`, it's cool because he'll only destroy the files and folders in his chroot.

Beyond Basic Containers

The above situation is still used quite a bit today, and there are many different variations on the basic chroot setup. Basic chroot is OK for some things, but like mentioned above it doesn't really seperate anything more than files. There was, and is, a clear need for something much more flexible and restrictive.

FreeBSD has the concept of BSD Jails which goes above and beyond by adding in things like disk, memory, and CPU quotas on FreeBSD systems. Solaris has Solaris Zones, introduced in 2004, which does pretty much the same thing as FreeBSD Jails but on Solaris systems. Many hosting companies run OpenVZ or Virtuozzo instead of full blown virtualization systems like Xen or KVM because they can cram many more containers onto a system than they can full virtual machines.

In 2008 LXC, or Linux Containers, was released for Linux. LXC was a joint effort between groups like Parrallels (who run Virtuozzo), IBM, Google, and many other individual developers to bring containization to the Linux kernel.

All of these more powerful containers came about because of the lack of quotas and security in chroot (not that I think this is a fault of chroot, chroot was not designed to handle those concerns). Containers are useful not only because they help protect users from the rest of the system, but also because they are generally much less resource intensive on the host machine.

Much like Virtualization though, running containers was not something that was easily done. Many times it meant setting up networking or quotas manually, or installing extras onto a system that most system admins did not deal with. There is a large amount of people that just do not know things like containers are even a thing.

Along Comes Docker

In 2013, dotCloud, a hosting company, released Docker. dotCloud used containers to deploy their customer's applications and handle scaling, and as the tooling around LXC was pretty bare bones they created what they needed. Docker was born from that need.

Docker originally ran on top of Linux Containers, and did for LXC what Vagrant did for virtual machines. By installing Docker onto a computer, you could easily build your own containers, package them up, distribute them, and create and destroy them with very few commands. You no longer needed to know much more than you needed for something like Vagrant to start playing around with containers.

Docker consists of a client and server. The Docker client allows you to issue commands to a server, which will then start, stop, destroy, or do other things with containers or images (basic 'installations' of containers). The Docker client also allows you to build images yourself.

The Docker server does the heavy lifting of setting up networking, interacting with some sort of container technology to run the containers, and all the management cruft you no longer have to worry about as a sysadmin or a developer. New technologies for Docker also now allow you to provision machines from a command-line client through Docker Machine, or orchestrate complex multi-container setups through Docker Compose.

New OSes have popped up as well, such as CoreOS which is a minimal host operating system for running containers. We are seeing very small Linux distributions, like Alpine Linux, which are being designed to be run as bases for containers. Like Vagrant, an ecosystem is starting to sprout up around containers and related technologies that we as developers can start to use.

Why We Should Care as Developers

From everything I've described, this sounds like something that will be really helpful for our system administrators, and it is. I'm not going to downplay the great ability to quickly deploy a known system very quickly and repeatibly across pretty much any system.

What's in it for us though?

Much like virtualization has helped remove the "It works on my machine" plague that has been the bane of web developers for years, containers takes that to another level. Docker containers are identical once they are built, so as long as everyone is using the same container base image they are 100% the same.

This differs slightly from virtual machines handled by configuration management like Puppet, which are prone to 'version creep' over time. For example, a project I started was set up to install PHP and Apache. Puppet correctly downloaded Apache and PHP and set up mod_php. This worked just fine up until Ubuntu switched from using mod_php to using the PHP Filter module in favor of mod_php. Since Vagrant only provisions a box generally at the first boot now, I was left with mod_php while new people on the project were getting the PHP Filter module, which does have a few differences. Yes, this can be fixed by reprovisioning the box, but unless you know to do that (and I was only aware of the change because I had to actually work on servers where this was an issue) most developers aren't going to notice this.

The other advantage is that containers are generally small. They (generally) contain only the needed files for a single process and not entire operating systems so their footprint is small. This means they are small to download and small to store.

The major advantage I find is that it allows developers to swap out pieces of their application as needed. As a PHP developer I can test my application using PHP 5.4, 5.5, 5.6, and 7.0 by swapping out a single portion of my setup. Want to run unit tests against multiple PHP versions without running different ones on your system? Containers, and Docker, will allow you to do that.

This also leads us to the fact that containerizing processes allows us to keep host systems "pure." You do not need to install PHP on your local system directly but can wrap it in an container. This keeps everything nicely packaged and you can quickly clean up systems without having all sorts of cruft in the host system. Couple that with the idea above and you have have multiple PHP versions running without conflict.

Throughout this book we'll explore setting up Docker and using it to our advantage as developers.

Chapter

2

Getting Started

Before we begin using Docker, we are going to need to install it. There will be a few caveats that we are going to discuss as we go through the installation because, unless you are on Linux, we're going to need some extra software to utilize Docker. This will create some extra issues down the road, but rest assured I'll keep you abreast of the more disasterous pitfalls that you may encounter, or various issues that might arise on non-Linux systems.

The installation is normally fairly easy no matter what OS you are going to use, so let's get cracking. We're going to install Docker Community Edition 17.03. I'll go over some basic installation, but you can always refer to *https://docs.docker.com/installation/* for anything special or other Operating Systems if you aren't using Windows, OSX, or Ubuntu.

Throughout this book, I'm going to be using Ubuntu for all of the examples because it not only gives us a full operating system to work with as a host, but it is also very easy to set up. There are smaller Linux distributions that are designed for running Docker, but we are more worried about development at this stage. Since we're using containers it doesn't really matter what the host OS is.

In 2016 Docker released Docker for Mac and Docker for Windows, which brings a much more native feel to working with containers on those operating systems. Everything that is discussed in this book should work fine on Mac or Windows. The few caveats that still exist are detailed in the section for those operating systems.

Docker Community Edition vs Docker Enterprise Edition

In March of 2017, Docker split the project into two versions - Docker Community Edition (CE) and Docker Enterprise Edition (EE). Functionally they are currently the same, with EE having a certification process for its releases and a different support plan than CE. You will not lose out on anything by using the CE edition of Docker, and as such we will be using it for this book. If you see references to Docker 1.13 in this book and online, that is essentially the same thing as Docker CE 17.03, as Docker changed the version numbers when releasing CE and EE.

Installing Docker

Ubuntu

Since Ubuntu ships with Long Term Release releases, I would recommend installing Ubuntu 16.04 and using that. The following instructions should work just fine for 14.04 or higher as well. Ubuntu does have Docker in it's repositories but it is generally out of date pretty quickly so we're going to use an apt repository from Docker that will keep us up-to-date. Head on over to *http://www.ubuntu.com/download/server* and download the 16.04 LTS Server ISO and install it like normal. If you'd like a GUI, grab the desktop version. Either one will work. I'll be using the Desktop version with the intent to deploy to Ubuntu 16.04 Server.

If you've never installed Ubuntu before, Ubuntu provides a quick tutorial on what to do. Server instructions can be found at *http://www.ubuntu.com/download/server/install-ubuntu-server* and Desktop instructions can be found at *http://www.ubuntu.com/download/desktop/install-ubuntu-desktop*.

There's a few commands we can run to set up the Docker repository. Open up a terminal and install Docker:

```
$ sudo -i
$ wget -qO- https://get.docker.com/ | sh
$ usermod -a -G docker [username]
$ exit
$ sg docker
```

Line 1 switches us to the root user to make things easier. Lines 2 runs a script that adds the repository to Ubuntu, updates apt, and install Docker for us. Line 3 sets up your user to use

Docker so that we do not have to be `root` all of the time, so replace [username] with your actual user you will use.

We can make sure that the Docker engine is working by running `docker -v` to see what version we are at:

```
$ docker -v
Docker version 1.13.0, build 49bf474
```

To make sure that the container system is working, we can run a small contaier.

```
$ docker run --rm hello-world
```

Ubuntu is all set up!

Windows 10 with Hyper-V

In 2016 Docker formally released a beta of Docker that runs on Windows, if your version of Windows includes Hyper-V. This includes Windows 10 Professional, Enterprise, and Education. If you have Home, you will need to upgrade to Windows 10 Professional to be able to use Hyper-V.

Head on over to the Docker Products[1] and download the package for Windows. The Docker for Windows package includes Docker Engine, Compose, Machine, and Swarm all ready to be installed, and the installer will also enable Hyper-V.

Launch the Installer. Accept the install agreement and let Docker install. Once the installer is done, make sure the 'Launch Docker' option is selected and finish the installation. That's it! See Figure 2-1.

Figure 2-1

Docker will start itself up. If Hyper-V is not installed, it will prompt you to install Hyper-V and then restart the PC (Figure 2-2). This may take a few additional moments, and your PC may restart a few times.

Once that is all finished you will have a 'Docker for Windows' icon on your desktop, and a Docker icon in your notification tray. As long as it is not red, you should be able to power up a Powershell window and run `docker -v` to make sure everything is working.

[1]   Docker Products: https://www.docker.com/products/docker#windows

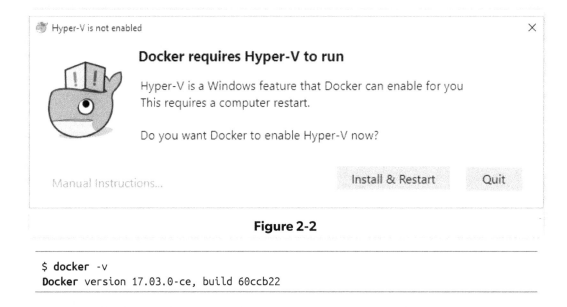

Figure 2-2

```
$ docker -v
Docker version 17.03.0-ce, build 60ccb22
```

Out of Memory Issues

On my test laptop, which has 4GB of RAM, I had to lower the VM memory usage all the way down to 1024 MB before it would run. I am not 100% sure why exactly, as I had more than 2 GB of available RAM at startup. If you get this error, right-click on the Docker icon in your notification tray, select 'Settings', and then 'Advanced.' Lower the Memory slider until you are able to start Docker.

pwd **In Code Samples**

Various shells in OSX and Linux allow you to specify a shortcut for the current directory by using pwd. *You will see this throughout the book. If you are on Windows, just replace* pwd *with the full path of the folder that you want.*

Windows 7/8/8.1

Docker does not support Docker Toolbox as well as Docker for Windows, and as such I cannot guarantee that everything in this book will work properly. Your mileage may vary.

Docker and Microsoft have released a version of Docker that runs under Hyper-V, but only for users of Windows 10 Professional or higher. For users of older versions, or lower versions of Windows 10, you will still need to download the Docker Toolbox[2], which will set everything up for us. The Toolbox includes the Docker client, Docker Machine, Docker Compose, Kitematic, and VirtualBox. It does not come with Docker Swarm.

Figure 2-3

Start up the installer. When you get to Figure 2-3 you can install VirtualBox and git if needed. I've already got them installed so I'll be skipping them but feel free to select those if needed. You should be good with the rest of the default options that the installer provides.

Since this changes your PATH, you will probably want to reboot Windows once it is all finished.

Figure 2-4

Once everything is all finished, there will be two new icons on your desktop or in your Start menu. Open up "Docker Quickstart Terminal." At this time Powershell and CMD support are somewhat lacking, so this terminal will be the best way to work with Docker. Once the terminal is up, run `docker -v` to see if you get a version number back.

```
$ docker -v
Docker version 1.9.0, build 76d6bc9
```

Since this is the first time you've opened up the terminal, you should have also seen it create a VM for your automatically. If you open VirtualBox you'll see a new 'default' VM, which is what Docker will use. Let's start a Docker container to make sure all that underlying software is working.

```
$ docker-machine run --rm hello-world
```

[2]   Docker Toolbox: https://www.docker.com/products/docker#windows

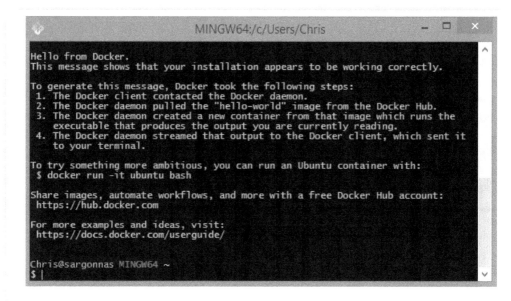

Figure 2-5

You should get a nice little output from Docker. If so, move on, you are all set!

Potential Problems

If the creation of the default VM seems to hang, you can cancel it with a CTRL+C. Open up a Powershell session and run the following commands:

```
$ docker-machine rm default
$ docker-machine create --driver=virtualbox default
```

This will create the virtual machine for you, and the Quickstart Terminal will work with it after that. For some reason either Docker or VirtualBox seems to hang sometimes creating the default VM inside the terminal, but creating it through Powershell will work. You may need to accept some UAC prompts from VirtualBox to create network connections during the creation process as well.

If you are trying to run any commands throughout this book and they are failing or not working correctly, it may be due to a big in the terminal. Try running the command by adding winpty to the beginning, like this:

```
$ winpty docker run --rm hello-world
```

winpty should be pre-installed in the Quickstart tutorial, and does a better job of passing commands correctly to the various Docker programs.

You may also want to use the Docker CLI bundled with Kitematic. Simply open Kitematic, and click on 'Docker CLI' in the lower left-hand corner. You'll get a custom Powershell that is configured to work with Docker. Most commands should work fine through here, and throughout the book I've noted any specific changes needed for Windows.

OSX 10.10

OSX Yosemite introduced a new virtualization layer, much like Windows Hyper-V, that allows much better and tigher control of virtual machines. Since Docker needs features that are part of the Linux kernel, Docker has introduced a new client and setup process that uses the native OSX 10.10 virtualization layer. This provides much better performance and ease-of-use compared to Docker Toolbox.

Head on over to the Docker Products[3] and download the package for OSX. Open up the Docker.img file and drag the Docker

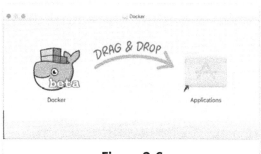

Figure 2-6

icon into your Application folder. Once it has copied, simply run the Docker application, and a Docker icon will appear in your notification area. See Figure 2-6.

Docker may ask for additional privileges, so allow it and type in your password. If it asks to update, feel free to let it go ahead and start the update.

Once that is all done, open up a Terminal session and run docker -v to make sure it all works:

```
$ docker -v
Docker version 1.11.2, build b9f10c9
```

OSX 10.9 and Earlier

Docker does not support Docker Toolbox as well as Docker for Mac, and as such I cannot guarantee that everything in this book will work properly. Your mileage may vary.

Since Docker is a native Linux application, we will need to run Docker through some sort of virtualization technology and inside of a Linux VM. OSX prior to 10.10, like many versions of Windows, does not have a native container or even virtualization system, so we are relegated

[3]   Docker Products: *https://www.docker.com/products/docker*

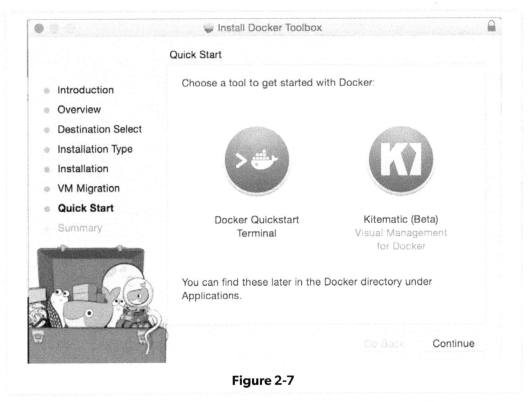

Figure 2-7

to using a tool like Virtualbox. Like Windows, Docker has the Docker Toolbox[4] available to download and set everything up for us. Go to the web, download the .PKG file, and open it up. There isn't much to change here other than if you want to send statistics to Docker, and which hard disk you want to install to, so the installation is painless.

Once the install is finished, click on the 'Docker Quickstart Terminal' icon in the installer to open that up. After it generates a new VM for us to use, you can make sure Docker is working by running `docker -v`:

```
$ docker -v
Docker version 1.9.0, build 76d6bc9
```

Finally, make sure that the container system is working by running the 'hello-world' container:

```
$ docker run --rm hello-world
```

You should end up with output like in Figure 2-8.

[4]   Docker Toolbox: https://www.docker.com/docker-toolbox

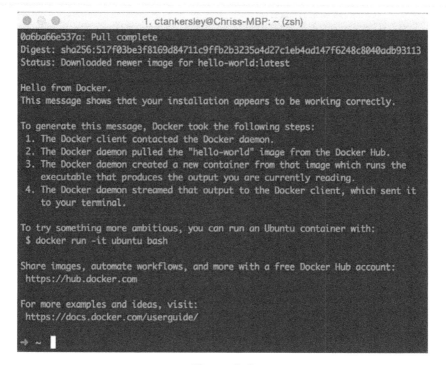

```
● ⌂ ●                    1. ctankersley@Chriss-MBP: ~ (zsh)
0a6ba66e537a: Pull complete
Digest: sha256:517f03be3f8169d84711c9ffb2b3235a4d27c1eb4ad147f6248c8040adb93113
Status: Downloaded newer image for hello-world:latest

Hello from Docker.
This message shows that your installation appears to be working correctly.

To generate this message, Docker took the following steps:
 1. The Docker client contacted the Docker daemon.
 2. The Docker daemon pulled the "hello-world" image from the Docker Hub.
 3. The Docker daemon created a new container from that image which runs the
    executable that produces the output you are currently reading.
 4. The Docker daemon streamed that output to the Docker client, which sent it
    to your terminal.

To try something more ambitious, you can run an Ubuntu container with:
 $ docker run -it ubuntu bash

Share images, automate workflows, and more with a free Docker Hub account:
 https://hub.docker.com

For more examples and ideas, visit:
 https://docs.docker.com/userguide/

→ ~ █
```

Figure 2-8

Potential Issues

One thing I have noticed, and has also been brought to my attention through readers, is that the \ character used throughout the code examples may not work in the Docker Quickstart Terminal for some reason. You may need to type the full command out instead of copy/pasting the examples.

You may also find that you have trouble mounting directories when using the Quickstart Terminal. You may be better off using a regular terminal session and accessing the Docker environment like so:

```
$ eval $(docker-machine env default)
```

If the above does not work to fix mounting directories, you may need to destroy and re-create the environment. You can rebuild the virtual machine by running:

```
$ docker-machine rm default
$ docker-machine create --driver virtualbox default
$ eval $(docker-machine env default)
```

Running Our First Container

Now that Docker is installed we can run our first container and start to examine what is going on. We'll run a basic shell inside of a container and see what we have. Run the following:

```
$ docker run -ti --rm ubuntu bash
```

You'll see some output similar to Figure 2-9.

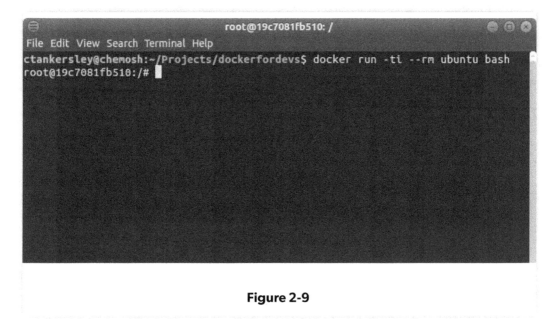

Figure 2-9

We told Docker to do was run a container based on Ubuntu, and inside of that run the command /bin/bash. Since we want a shell, run an interactive terminal (-ti), and when we are done with this container just delete it completely (--rm). Docker is smart enough to know that we do not have the Ubuntu image on our machine so it downloads it from the Docker Repository, which is an online collection of pre-built images, and sets it up on our system. Docker then runs the /bin/bash command inside that image and starts the container. We're left with the command line prompt that you can see at the bottom of Figure 2-3.

This container is a basic installation of Ubuntu so you get access to all of the GNU tools that Ubuntu ships with as well as all the normal commands you are used to. We can install new software in here using apt, download files off the internet, or do whatever we want. When we exit the shell provided, Docker will just delete everything, just like if you delete a virtual machine. It is as if the container never existed.

You'll notice that we are running as the root user. This is a root user that is specific to this container and does not impart any specific privileges to the container, or your user, at all. It is root only in the context of this container. If you want to add new users you can using the adduser command, just like regular Ubuntu, because everything that is part of a normal Ubuntu installation is here.

Now run ls /home. Assuming you haven't created any users this folder should be empty. Contrast this to your host machine that will have, at the very least, a directory for your normal user. This container isn't part of your host machine, it is it's own little world.

From a file and command perspective, we are working just like we did in the chroot. We are sectioned off into our own corner on the host machine and cannot see anything on the host machine.

Remember though, we aren't actually running Ubuntu inside the container though, because we aren't virtualizing and running an operating system inside a container. Exit the container by typing exit, and the container will disappear.

How Containers Work

For demonstration purposes, I'm running Ubuntu as the host machine. For the above example I'm using a Ubuntu-based image only because that was easy to do. We can run anything as a base image inside the container. Let's run a CentOS-based image:

```
$ docker run -ti --rm centos /bin/bash
```

The same thing happens as before - Docker realizes we do not have a CentOS image locally so it downloads it, unpacks it, and runs the /bin/bash command inside the newly downloaded image. The new container is 100% CentOS.

```
[root@90a244e62ee3 /]# cat /etc/redhat-release
CentOS Linux release 7.1.1503 (Core)
[root@90a244e62ee3 /]#
```

The important thing to keep in mind is that we are not running an operating system inside of the container. This second container is using an image of CentOS's files to execute commands, while the first example was using an image of Ubuntu's files to run commands.

When Unix executes a command, it starts to look around for things like other executables or library files. When we run these things normally on a machine, the OS tells it to look in specific folders, like /usr/lib/, for those files. The libraries are loaded and the program executes.

What a container does is tell a process to look somewhere else for those files. In our latter example, we run the command under the *context* of a CentOS machine, and feed the executable CentOS libaries. The executable then runs as if it is on CentOS, even though the host operating system is Ubuntu. If we go back to Chapter 1 and look at the basic chroot example, it is the same idea. The container is just looking in a different spot for files, much like Bob has to look in a different spot for /bin/bash.

The other thing that happens is that, by virtue of some things built into Docker and the Linux kernel, the process is segregated off from the other processes. For the most part, the only thing that can be seen inside the container are the processes that are booted up inside the container. If we run ps aux, we can see that only the bash process we started with, as well as our ps command show up:

```
[root@cc10adc8847c /]# ps aux
USER       PID %CPU %MEM    VSZ   RSS TTY      STAT START   TIME COMMAND
root         1  0.3  0.1  11752  2836 ?        Ss   00:51   0:00 /bin/bash
root        24  0.0  0.1  19772  2156 ?        R+   00:51   0:00 ps aux
```

If we run ps aux on the host machine we will see all the processes that are running on the host machine, including this bash process. If we spin up a second container, it will behave like the first container - it will only be able to see the processes that live inside of it.

Containers, normally, only contain the few processes needed to run a single service, and most of the time you should strive to run as few processes inside a single container as possible. As I've mentioned before, we aren't running full VMs. We're running processes, and by limiting the number of processes in a single container we will end up with greater flexibility. We will have more containers, but we will be able to swap them out as needed without too much distruption.

Containers, especially those running under Docker, will also have their own networking space. This means that the containers will generally be running on an internal network to the host. Docker provides wrappers for doing port forwarding to the containers which we will explore later in this book. Keep in mind though that the containers can reach the outside world, but the outside world may not necessarily be able to reach your containers.

Now, there's a whole lot of techincal things I'm leaving out here, but that's the gist of what we need to worry about. We're not running a full operating systems inside containers, just telling executables to use a different set of libraries and files to run against, and using various walls put up by the container system to make the process only see the files inside the image. Unlike a traditional virtual machine we are not booting up an entire kernel and all the related processes that will handle our application, we're simply running a program inside our PC and pointing it to a specific set of libraries to use (like using CentOS's libraries, or Debian's libraries).

Chapter

3

Working With Containers

In Chapter 2 I talked a bit about how containers work in a very general sense. While we booted up a container and dug around a bit there is still a lot of things that docker brings that will make our lives easier as developers. Let's begin to start to dive deeper into Docker and see how some of these things work.

Images

The heart of Docker comes in the form of Images. Think of these as base boxes from Vagrant. Images are pre-compiled sets of files that are built for either being built upon, like the Ubuntu or CentOS images we looked at earlier, or already set to run your code, like a PHP or nginx image. We can build these images ourselves or download them from the Docker Registry, which is an online hub of published and freely available images for you to use.

Images are not containers, but are the building blocks of containers. We download an image from the internet (or build one ourselves), and then we use the image to create a container. This is like the definition of a class and an object - classes define how something works, whereas an object is the thing that we actually use. We can create many containers from a single image, and as we go along we will do that quite a bit.

For Docker, we can see what images are currently installed by running `docker images`:

```
$ docker images
REPOSITORY          TAG             IMAGE ID        CREATED
ubuntu              latest          07f8e8c5e660    2 weeks ago
centos              latest          fd44297e2ddb    3 weeks ago
```

If you have been following along with the book, your image list will look much like the above list. We have two images, ubuntu and centos, with specific IDs and file sizes (which I've ommited to space). There's also an Image ID, which designates a specific hash of an image. There is also a TAG header. We can have multiple versions of images. For example, we currently have the latest version of the Ubuntu image on our machine, which at the time of this writing is 14.04.1. If we needed an older version, say 12.10, we can download that image using `docker pull` and supplying a tag name with our image name:

```
$ docker pull ubuntu:12.10
Pulling repository ubuntu
c5881f11ded9: Download complete
511136ea3c5a: Download complete
bac448df371d: Download complete
dfaad36d8984: Download complete
5796a7edb16b: Download complete
Status: Downloaded newer image for ubuntu:12.10
```

If you run `docker images` again you'll see two ubuntu entries, one set as 'latest' and the other as '12.10'.

The other thing to notice is that when we pull down an image, it will come in seperate chunks. Each image is comprised of a set of layers that can be shared between images. The Ubuntu 12.10 image has five layers. The reason for this is twofold - it allows images to be shared and cached between other images and it allows for quicker building of images. If we publish a PHP image based on Ubuntu 12.10, if someone already has the 12.10 image downloaded and they pull our PHP image, they will only need to grab the new layer where we added PHP. This reduces the amount of storage space needed for multiple images.

Images can come from the Docker Hub, the Docker Store, built from scratch using a Dockerfile, from a tarball of files, or from a private registry. We'll deal with the latter two options later, but I will generally see if there is an available image on the Docker Hub before building my own from scratch. Most of the time I will use a base image like Ubuntu and a Dockerfile and create my own with little work.

Controlling Containers

Docker helps add a bunch of nice wrappers around controlling and working with your containers. Docker allows us so run, start, stop, and destroy containers with a very easy syntax.

We've already seen how to initiate a container with docker run. Running a container kicks off the container by initializing the image, setting up the container, and starting the process inside of it. There are a couple of things we can do when we run the container.

```
$ docker run [options] IMAGE [command] [arguments]
```

[options] are options that will be used to configure the container. While there are a whole slew of them, we've been using three of them thus far. -t will start a TTY session inside the container that we can connect to, -i will run the container interactively meaning that it will be run like a normal process that we can control and interact with, and --rm tells the container to destroy itself as soon as it finishes. Putting this all together we end up with a /bin/bash process that we can interact with like a regular bash process, and it disappears when we type exit.

If we leave off --rm, the container will stop, but not delete itself, when we exit the container. If we have a process, for example a web server, we can leave that off so that our web server container doesn't disappear just because we stop it.

For containers that we do no need interactive access for, or want to run in the background, we can replace -ti with -d, which tells Docker to run the process in Daemon mode. This is like starting nginx or Apache using the service command, where the process kicks off and then gets shunted off to the background to run. It is there and available for us to interact with later, but it will not tie up our TTY session.

We have not used it yet, but one useful option we can pass is --name, which sets a specific name on a container. By default Docker will assign a random name to a container.

We'll look at more options as we go along.

IMAGE is the name of a Docker image we want to spawn. We've thus far looked at running containers based on Ubuntu and CentOS by virtue of the image names. You can search through the Docker Hub or build your own images and invoke them with the specified name.

[command] is the command we run inside the container itself. In our example so far, we've been using /bin/bash, so this runs /bin/bash inside the container. [arguments] are arguments for the [command] we are running, and are whatever arguments the [command] would normally take.

Now, once we have initiated a container with docker run, we can stop it with docker stop [name].

If you want to start a stopped container back up, you can do that with docker start [name]. You'll notice this does not use the image name, but the name that Docker (or you via --name) set for the container. Many of the commands we will use going forward can use the name of a container.

If you need to re-connect to a container after you have stopped it and started it, you can use `docker attach [name]`. You'll be pushed back into the container and can interact with it like you just started it with a run command.

Over time you will probably forget which containers are stopped, or what their names are. You can see a list of all the registered containers that are running by using `docker ps`. This lists only the running containers, so if you want to see all the containers you can use `docker ps -a` to list everything. You'll see quite a few stats for the containers and how they were originally started.

Once you are all finished with a container, you can delete it manually if you did not set it to auto-delete with `--rm`. You can use `docker rm [name]` to remove a stopped container. If you want to stop and delete at the same time, you can use `docker rm -f [name]` to do a force delete.

Container Data

Up until this point, our containers have just been disposable blobs of binaries that we can run and then destroy. We are limited to to whatever is inside the container when we pull it down (though, technically, we could download data into the container). Most of the stuff we will want to do, especially as developers, will require us to shove data inside containers, or be able to access the data inside of a container. Eventually we'll even want to be able to share data between containers.

Luckily, Docker has a concept called a Volume, which is a section of the image that is expected to have data that should be mounted into the image in some way. This is much like you plugging in a USB stick into your PC. The operating system knows there are entry points (be it in /Volumes if you are in OSX, or unused drive letters in Windows) where we will "plug in" or store data outside of the regular storage. We can tell Docker that a volume is either just a directory inside the image to persist or that the volume data will come from some external source such as another image or the host computer. Both have advantages and disadvantages, but we will discuss those later. For now, let's just start working with data.

For all of the Volume commands, we will start to use the -v flag to our `docker run` commands. You can even have multiple -v options to mount or create as many data volumes as we need. Some applications will need one, two, or many, and you may or may not mix host-mounted volumes with normal data volumes.

Mounting Host Directories

The quickest way to get files into a container is to just mount them into the container yourself. We can use -v `[hostpath]`:`[containerpath]` to mount a host directory or file into the container. Both `hostpath` and `containerpath` must be full paths. Relative ones will not work.

```
1.  $ mkdir testdir
2.  $ echo 'From the host' > testdir/text.txt
3.  $ docker run -ti --rm \
4.      -v `pwd`/testdir:/root \
5.      ubuntu /bin/bash
6.  root@eb10e0d46d2c:~# ls /root
7.  text.txt
8.  root@eb10e0d46d2c:~# cat /root/text.txt
9.  From the host
10. root@eb10e0d46d2c:~# echo 'From container' > /root/container.txt
11. root@eb10e0d46d2c:~# exit
12. exit
13. $ cat testdir/container.txt
14. From container
```

The example is pretty straight-forward, but we created a directory with a file inside of it, ran a container that mounted our testdir/ directory into /root/ inside the container, and were able to read and write from it. Pretty simple!

The major downside to mounting host volumes in a production environment is that the container now has access to your host file system. Anything written to the shared volumes is now accessible by the host. It also means that the file paths must 100% match between production and development, which isn't always guaranteed. I did cheat a bit by using pwd, which ends up creating a full path, but it still must be taken into consideration.

I do not consider this a huge issue in development, but it is something to keep in mind when moving into production.

Persisting Data with Data Volumes

This plays into more of a production environment, but normally you will need to persist data between containers, or even between rebuilds of containers. Under normal working conditions, you can start and stop a container and all your data *should* still be there. But what if you need to rebuild a container, or in the case of production deployments, swap a container our completely? Your data will be gone.

Docker has a built-in way to mark a container directory as a Volume. Anything that is flagged as a data volume will persist its data between rebuilds, restarts, and upgrades. There are a few different ways that you can use them, some old and some new, and each one depends on what you want to do.

The simpliest way is to just tell Docker that we have a data volume inside of our container, so we need to tell Docker about it. We can do that by passing -v [/path/in/container] with our create or run command, and Docker will section off that path as being persistant.

Windows Mounts

If you are a Windows user, the above will not work for you out of the box. You will need to go into the Docker Settings, click on 'Shared Drives', and enable which drives are allowed to be passed through to the VM. If you do not do this you will not be able to mount any host drives into Docker. Once this is enabled, you can pass a full Windows path without issue.

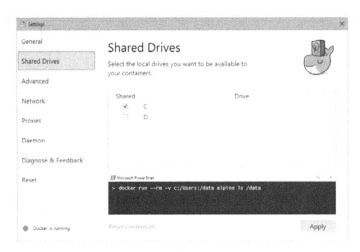

Docker for Windows Shared Drives

```
1. $ docker run -ti --name docker4devs \
2.      -v /root \
3.      ubuntu /bin/bash
4. root@5348b53817e3:/# cd /root
5. root@5348b53817e3:~# touch sample.txt
6. root@5348b53817e3:~# exit
7. exit
8. $ docker start docker4devs
9. docker4devs
10. $ docker attach docker4devs
11. root@5348b53817e3:/#
12. root@5348b53817e3:/# cd /root
13. root@5348b53817e3:~# ls
14. sample.txt
```

Volumes created in this way will stick around for as long as the container lives. If you destroy the container, you will destroy the underlying volume as well, so be careful. Volumes created like this are useful when you want to share the same volume across multiple images, but it requires the originating image to always be available.

What happens when we need to destroy a container to do something like upgrade to a new one? We need to create Named Data Volumes.

A Named Data Volume is simply a data volume that has it's own name, but is not directly part of a container. It can be used by any number of containers at the same time though, so they perfect places to store source code. As they are independent of containers they also make great places to store persistant data like database data.

To use them, we first need to create a named data volume. We do this with the `docker volume create [volume-name]` command.

```
$ docker volume create myvolume
myvolume
```

We can then reference it in `docker run` by using the colon syntax like with a host volume, but by passing the name of the volume instead.

```
1.  $ docker run -ti --rm \
2.      -v myvolume:/opt/data \
3.      ubuntu /bin/bash
4.  root@b38c6729f0bd:/# touch /opt/data/myfile.txt
5.  root@b38c6729f0bd:/# ls /opt/data/myfile.txt
6.  /opt/data/myfile.txt
7.  root@b38c6729f0bd:/# exit
8.  $ docker run --rm \
9.      -v myvolume:/opt/ \
10.     alpine ls /opt
11. myfile.txt
```

What we did was create an Ubuntu volume and mounted our named volume to /opt/data. Like with a host mounted volume, you can mount the named volumes anywhere in the container you want. We created a file called myfile.txt, and exited out. We then created an alpine image that mounted our volume to /opt and just listed out the directory contents. Inside was our file we created.

In both cases the containers were destroyed when we were finished. Without the named data volumes their data, including any specified volumes, would have been deleted as well. Since we used a named data volume the data persisted.

Named Data Volumes created this way also have the advantage of being able to use a pluggable architecture. By default volumes are created and hosted on the local Docker instance. You can also create volumes on things like GlusterFS or NFS to allow for volumes to be used across multiple nodes. Each driver has it's own set of additional options, so Google will be your friend with what is available and how to set them up.

You may also see people suggest using small containers based on alpine or busybox that do nothing but house volumes (in fact, earlier versions of this book did just that!). With the advent of named data volumes there is no need to do this anymore. Named Data Volumes have much better support with tools like Docker Compose and are much more flexible.

Networking

Part of working with containers is sectioning out concerns into different containers. Like with having to share data, we will have pieces of our application that will need to talk to each other. In a basic PHP application, we tend to have three layers to our application that need to talk together - the web server, the PHP process, and the database. While we could technically run them all inside of a container... we really should not. By splitting them into different containers we make it easier to modularize, but now we have to get them to talk together.

Docker automatically sets up and assigns IP addresses to each of the containers that run. For the commands that we have run thus far we haven't needed to care about this, but now we should start to dive into this aspect of Docker.

Legacy Linking

The original way of doing this with Docker, and still one of the easiest for when it just comes to messing around, is using Linking. This is supported only as a legacy method and is not generally a good fit for running in Production, but it does allow you to get your feet wet by starting to see how containers can talk back and forth over a network.

To show this off, we will create a few different containers. The first one we will create will be the PHP container, so that we can run a basic PHP script.

```
$ mkdir -p networking/html
$ echo '<?php phpinfo();' > networking/html/index.php
$ cd networking/
$ docker run -d --name app_php \
    -v `pwd`/html:/var/www/html \
    php:fpm
```

We generated a PHP container that is, for the moment, just going to mount a file from our host. We named it app_php just to make it a bit easier to work with in the following steps. If you run a docker ps, you'll also see that it is exposing port 9000 on it's IP address. This is not exposting it on the host IP, just the local address for the container. We can get this from the docker inspect command:

```
$ docker inspect \
    -f "{{.NetworkSettings.IPAddress}}" app_php
172.17.0.16
```

docker inspect by default will return a bunch of information about the container, but in this case we are only looking for the IP address. In my case, it returns 172.17.0.16 as the address. If I stop and start this container again, it will get a new IP address.

Windows and OSX users will find that this IP will not work for them, due to the virtual machine Docker is running inside. You will need to expose a port through the host, which is explained in the section named "Exposing Ports." This unfortunately is an area where Windows and OSX differ from the Linux implementation.

It's a pain to keep track of the IP address, but Docker as a built in way to somewhat discover and let containers know about the IP addresses of other containers. Let's spin up an nginx container and have it talk to PHP over port 9000. We'll need to modify the config for nginx to pass PHP requests to our PHP container. You can either use an nginx config you already have, or borrow the one from our sample app in docker/nginx/nginx.conf[1] (copy only the server {} block from that file, do not use the entire file). Put the configuration in a new file called default.conf, and we will mount it inside the container.

```
$ mkdir -p config
$ vi config/default.conf
$ docker run -d --name app_nginx \
    --volumes-from app_php \
    -v `pwd`/config/default.conf:/etc/nginx/conf.d/default.conf \
    --link app_php:phpserver \
    nginx
```

Well, that run command is a doozy. We're going to create a container that runs in the background named app_nginx. This container will mount the volumes from app_php inside of itself, much like we did before with the data containers. We're also going to mount a config file inside of it, which contains our configuration so we can have nginx talk to the php-fpm process that lives inside app_php. That is all pretty standard from the stuff I have gone over so far.

What is new is the --link [name]:[alias] option. This tells Docker to link the named container to the container you are creating, with the alias you specify. In this case, we are linking the app_php container with the name of phpserver. Inside the nginx container, we can use phpserver as the domain name for our PHP container!

If I grab the IP address for my app_nginx container using docker inspect, I can visit that IP and be greeted with a basic PHP info screen, like in Figure 3-1.

[1]   docker/nginx/nginx.conf:
    https://github.com/dragonmantank/dockerfordevs-app/blob/master/docker/nginx/nginx.conf

PHP Version 5.6.9

System	Linux 21b09859fe07 3.19.0-16-generic #16-Ubuntu SMP Thu Apr 30 16:09:58 UTC 2015 x86_64
Build Date	May 26 2015 17:31:50
Configure Command	'./configure' '--with-config-file-path=/usr/local/etc/php' '--with-config-file-scan-dir=/usr/local/etc/php/conf.d' '--enable-fpm' '--with-fpm-user=www-data' '--with-fpm-group=www-data' '--disable-cgi' '--enable-mysqlnd' '--with-curl' '--with-openssl' '--with-pcre' '--with-readline' '--with-recode' '--with-zlib'
Server API	FPM/FastCGI
Virtual Directory Support	disabled
Configuration File (php.ini) Path	/usr/local/etc/php
Loaded Configuration File	(none)
Scan this dir for additional .ini files	/usr/local/etc/php/conf.d
Additional .ini files parsed	(none)
PHP API	20131106
PHP Extension	20131226
Zend Extension	220131226
Zend Extension Build	API220131226,NTS
PHP Extension Build	API20131226,NTS
Debug Build	no
Thread Safety	disabled
Zend Signal Handling	disabled

Figure 3-1

Linking containers together has a nice side effect in newer versions of Docker, at least since v1.9. Docker will now track the IP correctly when linked containers come up and down. For example, if you have the above nginx and PHP containers, and turn PHP on and off, PHP will more than likely get a new IP. This IP will be given to the nginx container automatically. Older versions of Docker set the IP staticly when the container was built, so if PHP was at 10.10.0.2 when nginx came up, nginx would think PHP was at that IP until the nginx container was restarted.

Networking System

Newer versions of Docker contain the ability to create networks on demand and attach containers to them. This has many advantages over the old legacy linking, but the biggest is a built in service discovery. This allows containers to advertise themselves and make themselves available automatically to other containers, making scaling much more easy. You can also do things like set up multiple networks and attach containers to multiple networks, allowing you to segregate containers even further.

The first thing you will need to do is create a network. This is handled with the `docker network create` command.

```
$ docker network create appnetwork
96b1b7b68f82d2c7e46f576ebaa9213108e122cc594dbf3739d44acde18dbc1c
```

We can now attach containers to this with the --network parameter of docker run. We will attach an nginx container to it like this:

```
$ docker run -d --name app_nginx \
    --network appnetwork \
    nginx
45197e37a12e6710f55e5fa6213fe366777090689c77fd0debca08d8cbab764f
```

From the outset, this is not really any different than running a container on the default bridge network. Where this starts to get a bit more powerful is when we start to throw service discovery into the mix. Without setting up any specific network links, we can create a container that is on this same appnetwork network and ping the nginx container by it's name:

```
 1. $ docker run --rm -ti \
 2.     --network appnetwork \
 3.     alpine ping app_nginx
 4.
 5. PING app_nginx (172.18.0.2): 56 data bytes
 6. 64 bytes from 172.18.0.2: seq=0 ttl=64 time=0.099 ms
 7. 64 bytes from 172.18.0.2: seq=1 ttl=64 time=0.161 ms
 8. 64 bytes from 172.18.0.2: seq=2 ttl=64 time=0.092 ms
 9. ^C
10. --- app_nginx ping statistics ---
11. 3 packets transmitted, 3 packets received, 0% packet loss
12. round-trip min/avg/max = 0.092/0.117/0.161 ms
```

We created a small container based on the alpine image and ran ping app_nginx inside of it. Since it was also attached to the same network as the container named app_nginx, it was able to use the service discovery that is part of the Docker networking system. It made a DNS query to the DNS service that is attached to the network (run by Docker), and it returned the IP of the nginx container we created.

Docker allows you to assign aliases to containers using the --network-alias parameter, and query by those as well. Let's create two nginx containers with an alias of nginx and then see what happens if we try to ping nginx:

```
 1. $ docker run -d --name app_nginx1 \
 2.     --network-alias nginx \
 3.     --network appnetwork \
 4.     nginx
 5. 7daa96da99e4365d9ceb87891be27901a49dcad6df33b716ba1849f18f89f442
 6. $ docker run -d --name app_nginx2 \
 7.     --network-alias nginx \
 8.     --network appnetwork \
 9.     nginx
10. a4a962e8df234c77841c1136f7906e5e7e2ff94737b33b2967b3243c2a589622
11. $ docker run --rm -ti --network appnetwork alpine ping nginx
12. PING nginx (172.18.0.3): 56 data bytes
13. 64 bytes from 172.18.0.3: seq=0 ttl=64 time=0.088 ms
14. 64 bytes from 172.18.0.3: seq=1 ttl=64 time=0.098 ms
15. 64 bytes from 172.18.0.3: seq=2 ttl=64 time=0.124 ms
16. ^C
17. --- nginx ping statistics ---
18. 3 packets transmitted, 3 packets received, 0% packet loss
19. round-trip min/avg/max = 0.088/0.103/0.124 ms
```

Network Aliases allow us to mark multiple containers with the same alias. When we try and ping nginx, the DNS service in the network checks to see if there is a container with that name, and if not checks to see if there are any containers with that alias, and then returns one of the two containers that is assigned to that alias.

A slightly more real world scenario would be to set up an nginx container that talks to PHP FPM, and two FPM containers named fpm1 and fpm2. If they both share a network alias of phpfpm, then you can set the proxy_pass configuration in nginx to just phpfpm and nginx will be able to talk to both FPM containers. If one happens to drop off, the Docker service discovery will notice and stop sending traffic to it.

Docker Compose and Docker Swarm take heavy advantage of this, and we'll use this to our advantage later in the book. For now, look at creating a network using this new system instead of the legacy linking.

Exposing Ports

If we take a look at docker ps, we will see that the PHP container has an exposed port of 9000/tcp and our nginx container has 80/tcp and 443/tcp. The network system in Docker will only route traffic to ports that are specifically exposed, which we can control a run/create time, or via config. In the case of our two containers, the original configuration for them told Docker to expose these three ports.

If you need to expose a port manually, you can do so with the --expose=[ports] option. We could so something like the following:

```
$ docker run -ti --rm --expose=80 ubuntu /bin/bash
```

These ports are also only exposed to the container IP, not the host IP. Many times though we need want to expose a container to the outside world, such as in the case of our nginx container. We can use the -p [port] option to expose a port to the outside world.

```
$ docker run -ti --rm -p 80 ubuntu /bin/bash
```

This command will assign port 80 to a random port on the host. We can get what that host port is by running docker ps, or by running docker port [name] [port]. It is a random port though, and every time you expose it you will get a different host port.

If you want to assign an exposed port to a specific host port, you can do so with -p [host port]:[port], like with the following:

```
$ docker run -ti --rm -p 8080:80 ubuntu /bin/bash
```

This command will map port 8080 on the host to port 80 on the container. You can also specify the option as -p [IP]:[host port]:[port] if you need to bind the port to a specific IP address instead of 0.0.0.0. You will also need to be aware that, like normal, ports below 1024 are generally reserved and will need root access to bind to. If you want your nginx container to listen on port 80 of the host, you may need to run the docker create/run/start command as root instead of the normal user you use for Docker.

For Windows and OSX users, you will run into a few issues with port mapping due to Docker running inside of a virtual machine. Since Docker is running inside of a virtual machine, the basic -p [port] will not work for you to access the containers. The Virtual Machne is acting as the 'host' for the container, so you will need to use -p [host port]:[port] for any examples going forward.

Chapter

4

Creating Custom Images

Thus far we have been using images that have been created by either Docker or the community, through the Docker Hub[1]. These images were pre-built to handle a specific problem, such as running PHP, a web server, a database server, composer, etc. There is noting inherintly wrong with these image and, as far as general containers go, many will be good enough for what we need.

There will come a time where you either do not like how an image is set up, like the nginx or the MySQL container, or you need specialized setups. A good example is the MySQL image. An older version of it was nearly 2.5 gigabytes of space because the maintainer did not remove all the source code (the new version clocks in at a much more managable 350 megabytes, and Oracle actually offers even smaller variants). It was possible to create a much smaller one, so I would build a MySQL container myself.

[1]   Docker Hub: https://hub.docker.com/

You may also not trust an image, or be able to vet an image, and want to build one yourself. Some situations and industries require a more stringent auditing of software, so building your own images might be quicker than trying to explain to auditors that some third party image is safe.

The final reason for building your own image is for deployment. As I've mentioned before you can use host mounting for volumes, but that isn't scalable. Data volumes are not necessarily transferable between hosts. This leaves us with deploying our code inside of an image, which means we have to build a custom image.

Whatever the reason is, building a custom image is fairly easy. We'll define a list of steps that need to be applied, much like a bash script, and in the end we'll have an image.

Dockerfiles

Docker has a build system that uses a settings file called a 'Dockerfile.' The Dockerfile contains all the instructions needed for building an image. There are a few new commands we will use as well as commands for doing things like exposing ports and creating volumes.

For a quick reminder, we deploy containers (actual things that run) from images (copies of a system). So we will build a new Image, and then deploy that as a running Container.

The format of a Docker file is very simple. We have two basic things:

```
# Comments
INSTRUCTION arguments
```

Comments are denoted as a line that starts with a '#'. Comments help explain or remind you what you did six months ago. There's a good reason you needed that weirld sounding package.

Instructions are generally a single word, like RUN, ENV, or WORKDIR. We then supply an argument to the instruction. Instructions apply a change to the image that is being built. Remember how when we pull down an image there are multiple layers? Each instruction, or change, to a base image is a new layer. Docker supports the following instructions:

- ADD
- ARG
- CMD
- COPY
- ENTRYPOINT
- ENV
- EXPOSE
- FROM
- LABEL
- MAINTAINER

- STOPSIGNAL
- ONBUILD
- USER
- VOLUME
- WORKDIR

I will cover most of the common ones, further documentation can be found at
http://docs.docker.com/engine/reference/builder/.

Let's take a look at a very simple Dockerfile, in Figure 5-1.

Figure 5-1

```
1.  FROM ubuntu:14.04
2.  MAINTAINER Chris Tankersley <chris@ctankersley.com>
3.
4.  RUN apt-get update && apt-get install -y \
5.      nginx \
6.      && apt-get clean \
7.      && rm -rf /var/lib/apt/lists/*
8.
9.  COPY nginx.conf /etc/nginx/nginx.conf
10.
11. EXPOSE 80 443
12.
13. CMD ["nginx", "-g", "daemon off;"]
```

The Dockerfile is pretty easy to read, even if you have never looked at one before. If you are using the git repository for this book, you will find a copy of the docker file in docker/Dockerfile as well. I'll break it down line by line:

1. Build this container from the ubuntu:14.04 tag
2. Set a MAINTAINER flag, useful for when sharing on Docker Hub
3. Update the apt cache, and
4. and install the nginx package
5. and clean the apt cache
6. and remove and leftover lists to conserve space in the image
7. Copy the nginx.conf file from the local system into /etc/nginx/nginx.conf in the container
8. Expose two ports, 80 and 443
9. Start nginx using nginx -g daemon off;

Most Dockerfiles will contain `FROM`, `MAINTAINER`, `RUN`, `ADD`/`COPY`, `CMD`, `EXPOSE`, and `VOLUME`. If these instructions do not seem to fit the bill for what you want to do, check out the full Dockerfile documentation that I linked to above.

FROM

```
FROM <image>[:tag]
```

While we can create containers from an existing filesystem already on our computers, Docker has a mechanism for using an existing image as a base. In our example Dockerfile, we declared `FROM ubuntu:14.04`, meaning our custom image will use the `ubuntu:14.04` image as a base, and we will run all the commands from there.

Any Docker image can be used as a base image, and it is recommended to use one of the official base images to build your images. Docker recommends using the Debian image they supply as it is small, tightly controlled, and is a well-known distribution.

You must supply at least an image name (for example, 'ubuntu', 'php', 'debian'). If you do not, the latest tagged version will be used. If you need a specific tag, you can supply the full 'image:tag' format.

MAINTAINER

```
MAINTAINER <name> <email>
```

Sets the author field on generated images.

RUN

```
RUN <command>
RUN ["executable", "param1", "param2", ... ]
```

RUN runs a specific command inside the image that will be persisted. In our sample Dockerfile, we use RUN to update our apt cache and install nginx. Each RUN command in a Dockerfile will generate a new layer in your image, so it is recommended to try and combine as many like commands together. In our sample Dockerfile we put together four commands to install nginx and clean up apt, otherwise there would be four separate layers.

There are two forms for the RUN command, the first being the 'shell' form and the second being the 'exec' form. The 'shell' format will run the commands inside the shell that is running inside the container (for example, `/bin/sh`). If your container doesn't have a shell, or you need a command to be run exactly without variable substitution, use the 'exec' format.

ADD and COPY

```
COPY <local/file/path> </image/file/path>
ADD <http://domain.com/file> </image/file/path>
ADD <file.tar.gz> </image/path/>
```

ADD and COPY will move files into the image, but each one works slightly differently. Most of the time you are going to want to use COPY to move a file from outside of the image to inside.

COPY only works to move a file or folder into the image. It does not work on remote files.

ADD will move a file or folder into an image as well, but it supports remote URLs as well as auto-extraction of tar files. If you have a local files packaged into a .tar or .tar.gz you can use ADD to extract them inside of the container. The same works for remote tar files as well.

If you do not need the files after extraction (for example, you are downloading installer files and you no longer need them after install), then you are actually better off downloading the files using curl or wget and them deleting them when you are finished through the RUN instruction. ADDing an archive results in the files being committed to a layer, where downloading, running, and removing the files via RUN will not commit the layer until the entire RUN instruction is complete.

CMD

```
CMD ["executable", "param1", "param2", ... ]
CMD ["param1", "param2", ... ]
CMD <command> [arguments]
```

CMD is generally the command that will be run when you create and start a container if the user doesn't pass another one. In our sample Dockerfile, the CMD will start up nginx for us as a foreground application. If the "executable" portion of the CMD is left off you can use it in conjuction with an ENTRYPOINT instruction to provide default parameters.

The first form of CMD will work outside of a shell inside the container, while the third form will run inside /bin/sh -c. Most of the time you will probably want the first form.

If you pass php -S to the php-fpm container, we are overriding the default CMD. If you always want a command to run, use ENTRYPOINT.

ENTRYPOINT

```
ENTRYPOINT ["executable", "param1", "param2", ... ]
ENTRYPOINT <command> [arguments]
```

ENTRYPOINT configures a container to run as an executable. The 'composer/composer' image is an image that is built to run composer for us. This image isn't designed to execute anything except composer.

Using ENTRYPOINT versus CMD depends on what you want to to. For example, we can use the following command to start up the development server:

```
$ docker run \
    -d -v `pwd`:/var/www --name testphp \
    php:fpm php -S 0.0.0.0:80 -t /var/www/html
```

We can simplify our command down a bit by using the following in a Dockerfile:

```
FROM php
EXPOSE 80
ENTRYPOINT ["php", "-S", "0.0.0.0:80"]
CMD ["-t", "/var/www/html"]
```

This sets the command that will always execute to php -S 0.0.0.0:80, and we can override the -t /var/www/html if we want to. If we build this Dockerfile as 'phpdevserver', we can run it like the following:

```
$ docker run -d -v `pwd`:/var/www phpdevserver
// Or to override the path we are watching
$ docker run -d -v `pwd`:/opt/app phpdevserver -t /opt/app/html
```

This does lock our container down to always running a specific command, but still allow us to modify the command slightly as we need.

EXPOSE

```
EXPOSE <port1> [port2 port3 ...]
```

This exposes a port that the container will listen on. This does *not* bind the port to the host machine, just simply exposes it through the containers IP address. If you need to bind an EXPOSEd port to the outside world via the host, you will need to use the -p parameter on docker run.

You simply specify a list of ports, separated by a space, and Docker will open them on the container.

VOLUME

```
VOLUME ["/path/inside/image"]
VOLUME /path/inside/image
```

This creates a placeholder for a volume inside of an image. This is useful for things like automatically creating a volume to hold log files, as containers that are started will automatically set these volumes up without you having to specify them via the -v flag on docker run.

You can use a combination of the VOLUME instruction with -v as well, so you are not limited to just the volumes that the Dockerfile creates.

Building a Custom Image

The Dockerfile is just a recipe though, it doesn't do anything special by itself. We use it to build an image using the docker build command. We can build this with the following command:

```
$ docker build -t customnginx .
```

We are adding a customer tag with -t customnginx so that it is easier to find and re-use this image. We then supply it a directory where the Dockerfile exists. By using . we are telling Docker to build in the current directory. You should see some output like below:

```
1.  $ docker build -t customnginx .
2.  Sending build context to Docker daemon 3.584 kB
3.  Step 1 : FROM ubuntu:14.04
4.  14.04: Pulling from library/ubuntu
5.
6.  Digest: sha256:d4b37c2fd31b0693e9e446e56a175d142b098b5056be2083fb4afe5f97c78fe5
7.  Status: Downloaded newer image for ubuntu:14.04
8.   ---> 1d073211c498
9.  Step 2 : MAINTAINER Chris Tankersley <chris@ctankersley.com>
10.  ---> Running in 60f06fa69695
11.  ---> 1d1bbaca5635
12. Removing intermediate container 60f06fa69695
13. Step 3 : RUN apt-get update && apt-get upgrade
```

There will be a lot more as it runs through the commands. Each Step is a layer in the total image. It may take a few minutes to build the image, but once it is finished you should be able to run docker images and see our custom image in there.

```
$ docker images
REPOSITORY          TAG             IMAGE ID          CREATED
customnginx         latest          2af5f1cfc29e      6 seconds ago
```

We can now use this image instead of the generic nginx container we were using before, and no longer have to mount our custom nginx.conf file through the command line.

```
$ docker rm -f nginx
$ docker run \
    --name nginx --link phpserver:phpserver -v `pwd`:/var/www -d \
    customnginx
```

Last chapter I talked about how important it was to be able to seperate out the components of your application. Here we started with a generic nginx container by using the nginx image, and we've swapped that out using our custom customnginx image without having to touch the PHP or the MySQL layer.

Chapter

5

Docker Compose

Up to this point in the book, we have mostly been looking at working with a single container. Manually configuring containers with `docker run`, creating networks with `docker network`, and managing all of that by hand is a very possible way of setting up your application, but Docker Compose allows us to specify all of that configuration through a simple file. Docker Compose will also handle managing all of the containers and setup through a set of basic commands.

Docker Compose is based on an older orchastration tool called 'Fig[1]'. The idea of Fig was to make it easy to define full-stack applications in a single configuration file yet still help maintain the "one process per application" design of Docker. Docker Compose extends this idea into an official tool that will allow you to define how applications are structured into Docker containers while bundling everything up into an easy-to-deploy configuration.

Compose will read a file and build the necessary `docker run` commands that we need, in the proper order, to deploy our application. Since you should understand how all of this works from a manual process, Compose will just make the work you have done thus far much, much easier.

[1]   Fig: http://www.fig.sh/

Docker Compose now works with Docker Swarm and will allow you to deploy your application to a remote Swarm, and will even allow you to deploy to a single remote box running Docker through Docker Machine. I would highly looking at Docker Compose when starting a new project, even if you only need a few containers. If nothing else it makes it much easier to share configuration and setup between all of the team members.

`docker-compose.yml`

Compose starts with a `docker-compose.yml` file which contains all of the information on each container inside of an application, and how it should be configured. Since this books is mainly geared toward PHP developers, we can break down a three teir PHP stack into nginx, PHP FPM, and MySQL. We also want MySQL's data to be persistant between runs and upgrades so we will designate a volume to store the databases in. Docker Compose will also create a network for us so all of the containers can talk to each other.

Any of the things that we can configure through `docker run` can be configured through the `docker-compose.yml` file.

```
1.  version: '3'
2.
3.  volumes:
4.    mysqldata:
5.
6.  services:
7.    phpserver:
8.      build: ./docker/php
9.      volumes:
10.       - ./:/var/www/
11.
12.    mysqlserver:
13.      image: mysql
14.      environment:
15.        MYSQL_DATABASE: dockerfordevs
16.        MYSQL_ROOT_PASSWORD: docker
17.      volumes:
18.        - mysqldata:/var/lib/mysql
19.
20.    nginx:
21.      build: ./docker/nginx
22.      ports:
23.        - "80:80"
24.        - "443:443"
```

By knowing how to use `docker run` to start up our application containers, the `docker-compose.yml` file is fairly straight-forward. We give each container a specific name, such as 'mysqlserver', 'nginx', or 'phpserver'. Each container can use a Dockerfile build to build from, like we do with our 'nginx' container, or just point to an image that we can modify through parameters, like our 'mysql' and 'phpserver' containers. The networking system will assign aliases to the containers using the service names we provided, so they can talk to each other over their network. We can specify the volumes that we want to both preserve, in the case of the 'mysqlserver' container, as well as the directories that we want to mount, in the case of the 'phpserver' container.

The full specification for the Compose file can be found at _https://docs.docker.com/compose/compose-file/_.

The `version` Key

The newest versions of Docker Compose will support up to a version 3 of the configuration file. Through the lifetime of Compose it has gone from just a quick orchestration tool to a full blown deployment and management tool, and with that we have three versions of the `docker-compose.yml` file format.

Version 1 was the original. If you were an early reader of this book all of the examples were in this version 1 format. It basically allows you to specify the container and their options, and worked through the legacy linking network. Scaling up and down, while supported, did not work very well.

Version 2 introduced the idea of services and volumes. It also started creating networks for the containers and utilized the new Docker networking layer. The format was basically the same, though containers now lived under the `services:` key. You could specify network and volume options as well.

Version 3 is the newest version and is Version 2 but with deployment options. You can specify how many instances of a service to run and where they should run.

If you are running Docker 1.13 or higher, I would suggest just going straight to version 3 even if you are not using it for deployment. Using version 3 will just make it easier if you go down that path, and all of the version 2 options are supported.

If you find other tutorials online that are not using version 2 or 3 as their syntax, I would not use that tutorial. While they will technically work because Compose will use the older versions, they will not give you all of the current information.

Running Containers

To boot up a Compose configuration, go into the directory containing the
`docker-compose.yml` file. We can boot the entire configuration using `docker-compose -d up`, and
we should see something similar to the following:

```
1. $ docker-compose up -d
2. Creating dockerfordevs_mysqlserver_1...
3. Creating dockerfordevs_phpserver_1...
4. Creating dockerfordevs_nginx_1...
5.
6. $ docker-compose ps
7.          Name                    Command           State         Ports
8. ----------------------------------------------------------------------------
9. dockerfordevs_mysqlserver_1    /entrypoint.sh mysqld   Up       3306/tcp
10. dockerfordevs_nginx_1         nginx -g daemon off;    Up       0.0.0.0:443->443/tcp,
0.0.0.0:80->80/tcp
11. dockerfordevs_phpserver_1     php-fpm                 Up       9000/tcp
```

Each container in the configuration gets appended with the name of the project, which is
picked up from the folder that the configuration is in or specified with -p, the name of the actual
container, and a number. The number is important because we can actually scale up the number
of each container as needed, which we will take a look at in a bit.

Once all of the containers have been created you can work with them like normal. We port
forwarded port 80 to localhost, so you can visit http://localhost and visit your site. We specified host volume mounting under the volumes key which means we can edit the files on the host
PC using our normal tools.

If we change anything in the `docker-compose.yml` file, we can just re-run
`docker-compose up -d` and it will only rebuild the containers that changed. If you make a chance
to a Dockerfile, you can add `--build` and Compose will rebuild the containers for you.

When you are all finished working on a project, you can turn everything off with a simple
`docker-compose stop`. The containers will stop as if you had individually stopped them with
`docker stop [name]`. If you want to completely delete the containers, `docker-compose rm` will
remove all of the containers that are listed in the `docker-compose.yml` file.

Testing Scaling

If you want to test how well your app handles scaling up and down in a controlled environment, Compose will allow you to scale services as you need using the `docker-compose scale`
command. Just specify the service name and how many you want to be running.

```
$ docker-compose scale phpserver=4
Creating dockerfordevs_phpserver_2...
Creating dockerfordevs_phpserver_3...
Creating dockerfordevs_phpserver_4...
Starting dockerfordevs_phpserver_2...
Starting dockerfordevs_phpserver_3...
Starting dockerfordevs_phpserver_4...
```

In the case of nginx, we can set the proxy_pass to phpserver and Docker will handle routing traffic to one of the four FPM containers we now have. Since the service discovery is all done through DNS everything should continue to work even if one of more of these new containers drops out. This scaling does work differently than Docker Swarm's scaling as Compose cannot handle additional host port allocation.

If dockerfordevs_nginx_1 is assigned to port 80 on the host, like we specified, when we attempt to scale it up the newly created nginx containers will fail to run as they will also try to allocate port 80 on the host. Our PHP containers can scale just fine as they are not using ports exposed through the host.

Docker Swarm does some network routing to handle this, so Compose can only be used to scale up containers that do not have host port mappings.

Chapter 6

Containerizing Your Application

Now that we understand how containers work, and a bit about how we can work with the containers, lets start to take an application and convert it over to using Docker. At first we will do this all by hand, but then we will begin to look at other tools to make it slightly easier. We will start with 'stock' images and move to to fine tuning everything.

We are going to take a simple PHP application, written in Zend Expressive, and chop it up into containers. Normally, an application will live inside of a single virtual machine that is managed by something like Vagrant. Having all of the parts of the application running in a single Docker container, while doable, presents some unique issues. Docker is a single-process-per-container ideal, so having our PHP, web server, and database all inside a single container is not the best. We will split up things based on spheres of influence.

First, go to *https://github.com/learningcontainers/dockerfordevs* and clone the repository to your local machine. For the purposes of demonstration, don't do anything else. You can either copy the text from this section as we go along, or switch to the various branches that helps fill in things for you.

Setting up a Compose file

Branch for this section: `git checkout containerize/compose-file`

Our application is a pretty bog standard PHP application, which means we will need a web server that runs PHP, and a database server. Throughout the book I've been using nginx, PHP-FPM, and MySQL, so we will continue to do that. We will also want our database to survive rebuilding the MySQL container if needed, so we will need to set up a data volumef for it. As we are working on this live, host mounting will also be needed. Of course, all of the containers will need to talk to each other.

I will start with some base images, so I have selected the `nginx`, `php:fpm-alpine`, and `mysql/mysql-server` containers. The latter two are a bit smaller versions of the their installations and should work just fine. I also know that we'll need some nginx configuration to point nginx back to PHP, so we will host mount those files for the moment. I will also forward `localhost:8080` to nginx.

The following `docker-compose.yml` file should suffice for the moment.

```
 1.  version: '3'
 2.
 3.  volumes:
 4.    mysqldata:
 5.      driver: local
 6.
 7.  services:
 8.    nginx:
 9.      image: nginx
10.      volumes:
11.        - ./public:/var/www/public
12.        - ./docker/nginx/default.conf:/etc/nginx/conf.d/default.conf
13.      ports:
14.        - 8080:80
15.
16.    phpserver:
17.      image: php:fpm-alpine
18.      volumes:
19.        - ./:/var/www
20.
21.    mysqlserver:
22.      image: mysql/mysql-server
```

```
23.      environment:
24.        MYSQL_DATABASE: dockerfordevs
25.        MYSQL_USER: dockerfordevs
26.        MYSQL_PASSWORD: 'd1ffp@ssword'
27.        MYSQL_ROOT_PASSWORD: 's3curep@assword'
28.      volumes:
29.        - mysqldata:/var/lib/mysql
```

With that configuration, a simple docker-compose up -d will start all of our containers.

For a sanity check, we can hit http://localhost:8080 and we should get an error about vendor/autoload.php not being found, as we haven't run Composer yet. If you get a screen similar to Figure 6-1, we are off to a good start.

We'll mess with the docker-compose.yml file some more in a bit, but at the very least so far we have nginx talking to PHP.

Figure 6-1

Running Composer

Branch for this section: `git checkout containerize/running-composer`

No modern PHP application is complete without using Composer, and this trivial application is no different. We need to run Composer to pull down our dependencies, but we can leverage Docker and it's containers to do this for us. For the moment, head over to *https://getcomposer.org/composer.phar* and download the latest Composer PHAR file. While we could go through the whole host mounting procedure with a `docker run` command, Compose exposes it's own `run` command to execute commands using Compose configuration.

The format of the `docker-compose run` command is:

```
docker-compose run [compose options] [service name] [command]
```

The options for Compose allow you to change a few basic things like the Entrypoint (command that is called by default), the working directory, change environment variables, and a few other things. We then specify a service name, *not* an image name, from the `docker-compose.yml` file, and finally the command to run inside the container.

This is doubly cool because all of the PHP containers from Docker Hub have the PHP CLI installed in them. We can then install our dependencies like this:

```
$ wget https://getcomposer.org/composer.phar
$ docker-compose run --rm \
    --entrypoint php \
    -w /var/www \
    phpserver \
    composer.phar install
```

Since we are using the PHP FPM container, it's default entry point is a small script it uses to bootstrap FPM. We use `--entrypoint php` to tell the image to instead run the `php` binary. The working directory of the image is also defaulted to `/`, so we change it with `-w /var/www` to run at the same path we mounted our PHP files. We then specify that we want to use the `phpserver` service, as defined in the `docker-compose.yml` file. This starts the container with all the options specified for that service, which in our case is just the host mounting. We then finish up the command with `composer.phar install`, which is tacked on to our new entrypoint. PHP runs the PHAR file, and our dependencies are installed.

If you visit the site now, you should get an "Oops!" page. This is another intended error, as this application is meant to use only a few routes. The page should look like Figure 6-2. If it does, feel free to move onto the next section.

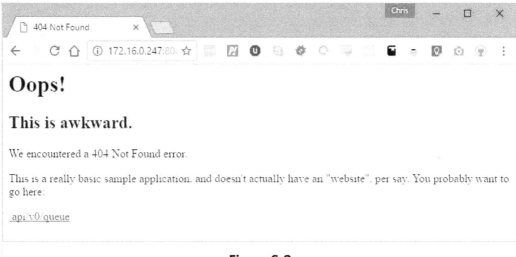

Figure 6-2

PHP Extensions

Branch for this section: `git checkout containerize/php-dockerfile`

While that Oops page is nice, the meat and potatos of this app is in the small API that is exposes. If you click the link at the bottom of the Oops page it should take you to `http://localhost:8080/api/v0/queue`. That will expose another problem we have though, as seen in Figure 6-3.

The error is a bit cryptic, but essentially we haven't provided any configuration for our database. That is not really Docker's fault, so lets fix that. Copy `config/autoload/local.php.dist` to `config/autoload/local.php`. If you want, change the `user` and `password` to the custom user we set up in the `docker-compose.yml` file, but the configuration should match the root user's password.

Refresh the page and we should get a new error, as in Figure 6-3. Remember, new errors are at least progress. In this case, PHP is complaining that there isn't any MySQL PDO driver available. That is a bit of a big problem, and one you will run into very quickly with the official PHP images. To keep their size down they ship with almost no extensions installed or enabled. That means we will need to create our own Dockerfile for the PHP server.

If you have not checked out the chapter on creating your own images yet, I would do so now. Creating a custom Dockerfile is fairly easy for PHP as the images have a few helper commands.

The first thing we need to do is create a new Dockerfile for PHP. Create a new file at `docker/php/Dockerfile`, and fill it in with the following:

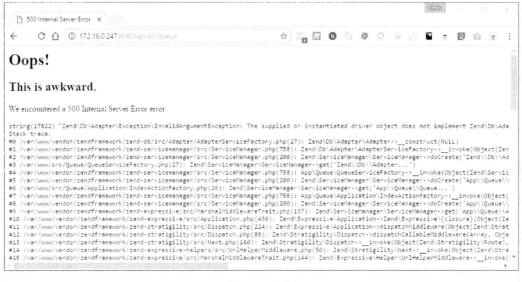

Figure 6-3

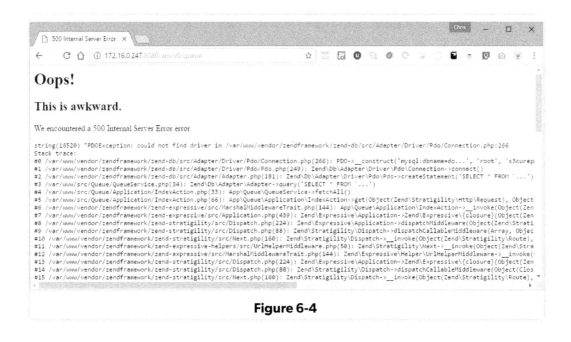

Figure 6-4

```
FROM php:fpm-alpine
```

```
RUN docker-php-ext-install pdo_mysql
```

docker-php-ext-install is a helper program that will download the MySQL PDO extension, compile it, and install it for us. That's all we really need for the Dockerfile, but we will not be hand building this image. We will let Compose do that for us.

In docker-compose.yml, change the phpserver service to the following:

```
phpserver:
  build: docker/php
  volumes:
    - ./:/var/wwww
```

We are swapping out the image key for a new build key. The build key tells Compose to build an image from a customer Dockerfile at the proposed path. Save the file, and run docker-compose up -d --build. We specify --build because we changed the Dockerfile, and we want Compose to rebuild the image.

```
$ docker-compose up -d --build
Building phpserver
[Whole bunch of build text]
Successfully built 52c6619524ee
dockerfordevsapp_nginx_1 is up-to-date
Recreating dockerfordevsapp_phpserver_1
dockerfordevsapp_mysqlserver_1 is up-to-date
```

If we refresh one more time, we should get an error about a missing base table or view, like Figure 6-4. Progress! This means that PHP is now talking to MySQL, but MySQL says the table we are looking for (queues), does not exist. We can fix that in a moment.

Development Tools

Branch for this section: git checkout containerize/dev-tools

We need to start venturing into tooling territory. Our application uses a database migration tool called Phinx[1], and one thing we will need to do quite regularly is to run database migrations. I've steathily left out that package from composer.json so we need to add that in. Before we do that we can talk about containerizing applications and tools.

While we will not run into it in this project, there are many PHP packages that require specific PHP extensions. While we have taken care of that with our phpserver service using

[1]   Phinx: https://phinx.org/

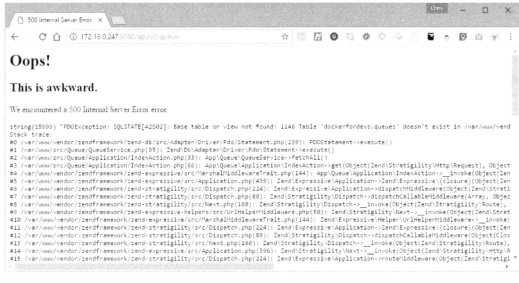

Figure 6-5

a custom image we can get rid of the necessity to manually download the Composer PHAR ourselves. Create a new file called `docker-compose.dev.yml` and add the following text to it:

```
1.  version: '3'
2.
3.  services:
4.    composer:
5.      build: docker/composer
6.      volumes:
7.        - ./:/var/wwww
8.      depends_on:
9.        - phpserver
```

We are going to create a new service named `composer` that we are going to use to run all of our composer commands. It will have a custom Dockerfile so that it can have all of the PHP dependencies that our packages may need.

What do we need in our customer Dockerfile for this? Create a new file at `docker/composer/Dockerfile` and put in the following:

```
FROM dockerfordevs_phpserver

RUN [install instructions from getcomposer.org]

WORKINGDIR /var/www
ENTRYPOINT ["/bin/composer"]
CMD ["install"]
```

I have truncated the RUN command as the install commands are pretty long, so you probably want to check out the repository version of this. Anyway, we are going to use the image that is build from our our PHP server service as a base, run the composer install commands, and set an entrypoint for /bin/composer and a default command of install.

But wait, how are we using dockerfordevs_phpserver as an image? Docker Compose is building a custom image for our phpserver service, and it has a consistent naming scheme (projectname_service). Since we designated the composer service to depend on the phpserver service, it will build the phpserver image first. Even though this is technically an FPM image, it contains the PHP CLI binary as well so we just override the entrypoint and command to force it to run composer. Now we have one file to maintain for our PHP dependencies (docker/php/Dockerfile) and a Composer image that we can use for our tool.

I had you put the configuration in a separate file for a reason as well. We can stack up configuration files to build configuration that is specific to our environment. We won't need to run Composer in a "production" environment, so we move it to a separate file. We can now build all of our images like this:

```
$ docker-compose \
    -f docker-compose.yml \
    -f docker-compose.dev.yml \
    up -d --build
```

You should see Docker build the phpserver image and then our new composer image. If you run docker-compose ps you will not see the composer container running, and for good reason. It defaults to running the install command, which we have done already. It finished it's task and exited. We can still use it's configuration for running commands though. We need to add the robmorgan/phinx package, so we can do that like this now:

```
$ docker-compose \
    -f docker-compose.yml \
    -f docker-compose.dev.yml \
    run --rm composer \
    require "robmorgan/phinx"
```

The command is a bit longer with the additional file, but we are starting to make our development tools part of the application. You can extend this to larger and more complicated tasks like SASS compilation and the multitude of node tools you need for Javascript. By creating custom command containers you make it much easier to get up and running with a project as well keep control of what versions of tools are used per project.

Now we need to run the database migrations. You can either create a dev tool to do this, or run the command inside the phpserver image that was created for us. It is up to you. For the purposes of this I will just use a command similiar to the beginnnig of this chapter when we ran Composer the first time:

```
 1. $ docker-compose run --rm \
 2.     --entrypoint php \
 3.     -w /var/www \
 4.     phpserver \
 5.     vendor/bin/phinx init
 6. Phinx by Rob Morgan - https://phinx.org. 0.7.1
 7.
 8. created ./phinx.yml
 9. $ vim phinx.yml
10.
11. // Edit the migrations as below
12.
13. $ docker-compose run --rm \
14.     --entrypoint php \
15.     -w /var/www \
16.     phpserver \
17.     vendor/bin/phinx migrate
18. [Migration should run]
```

The git repository has the file already prefilled for you, but there are a few changes we need to make if you are following along by hand. Inside the phinx.yml file you will need to point the migrations: key to our database migrations, which is %%PHINX_CONFIG_DIR%%/data/migrations. You will also need to go down to the development section and change the host, name, user, and pass keys to match our docker-compose.yml file. host will be mysqlserver as we will let the Docker DNS service handle the routing.

If everything was filled in correctly the migrate command should run our database migrations. If we refresh our page we should get real API output like Figure 6-6!

And with that, we containerized our application.

Figure 6-6

Thinking About Architecture

I will be honest, we didn't do much here but put into place many of the commands we talked about in Chapter 3 into a Docker Compose file. What I wanted to show is that when we start to look at using Docker, we need to start thinking about the architecture of our applications. Many of us are not building small little single-script command line scripts, though we did use Composer inside of a Docker container to show that you can do that. Our applications our multi-layered and are more than just PHP scripts. They are web servers, script interpreters, database servers, queue engines, third party APIs, data stores… we can break down many of our applications into smaller chunks.

None of the above should have required any code changes at all. If you take a look at your application, we did not change any of the source code for our application. We set up a config file to point the appropriate database server hostname, but we have to do that in any environment. We did not change any of the mainline application code to put it into Docker.

While you can use raw `docker run` commands to accomplish the same thing, and in fact older text of this book hand you do just that, Docker Compose is the go-to tool now for setting up environments. It is much more dependable and deployable to team members. In the next few chapters we'll take a look at deploying using this configuration.

There are many images out there, and even many tutorials, that say it's not a bad thing to run multiple processes in a single container. Many images for PHP contain nginx or Apache inside of them. The downside to running multiple processes inside one container is flexibility. If I wanted to use multiple versions of PHP, or swap between database engines, or test against nginx and Apache, I would have to build custom containers for all of those situations. By breaking it out I can quickly swap them with little work.

That's not to say that custom containers are bad. The PHP-based containers we use will almost always need custom images to handle our dependency needs. There is not a container that handles every situation.

Chapter

7

Docker Machine

Docker, like just about any software, can be installed from a package. If you followed along with this book thus far you've either installed Docker through a package repository or by using an software download from docker.com. That's all pretty standard. Doing it this way doesn't scale well though. Docker has a solution for that in the form of Docker Machine, an additional bit of software that makes provisioning machines for Docker much easier.

If you have installed Docker Toolbox, Docker for Mac, or Docker for Windows, you are all set to use it. If not, head over to *https://github.com/docker/machine/releases/* and follow the instructions for setting it up. More than likely you will just download a zip file and extract it to the appropriate places.

Once installed we can use to to create a brand new, pre-configured machine that we can use directly or join as part of a Docker Swarm.

To create a new machine we will run the `docker-machine create` command and specify the appropriate options for the driver. Docker Machine supports many different providers for creating hosts, such as:

- Amazon EC2
- Digital Ocean
- Microsoft Azure
- VMWare
- VirtualBox
- HyperV

and many, many others. Each service requires their own additional confguration options, but for the purposes of demonstration I'll set up a new machine on Digital Ocean[1].

We will need to get an API access token from Digital Ocean. If you do not have an account head over to https://digitalocean.com[2] and sign up for a new account. Once you are in your account click on 'API', and then generate a new token. We'll need this to create a new machine. Once we have an API token we can create the machine using the following command:

```
$ docker-machine create \
   --driver digitalocean \
   --digitalocean-access-token [token] \
   dockerfordevs

$ docker-machine ls
NAME            ACTIVE   DRIVER        STATE     URL
dockerfordevs   -        digitalocean  Running   tcp://XXX.XXX.XXX.XXX:2376
```

Each driver for Docker Machine has wildly different settings and parameters. For example, Digital Ocean's driver does not allow us to specify a RAM or hard drive size but instead allows you to specify a droplet size, where Hyper-V has you specify a virtual switch it must attach to. You will want to run docker-machine create --driver [driver] -h to get a list of the specific options for your driver.

The create command just creates the remote machine, but does not point our terminal at it. For that, we will need to change a few environment variables to point to the remote machine. We can get that command with docker-machine env dockerfordevs:

```
$ docker-machine env dockerfordevs
export DOCKER_TLS_VERIFY="1"
export DOCKER_HOST="tcp://XXX.XXX.XXX.XXX:2376"
export DOCKER_CERT_PATH="/home/user/.docker/machine/machines/dockerfordevs"
export DOCKER_MACHINE_NAME="dockerfordevs"
# Run this command to configure your shell:
# eval "$(docker-machine env dockerfordevs)"
```

[1]   Digital Ocean: https://www.digitalocean.com/?refcode=142755e4323e
[2]   https://digitalocean.com: https://www.digitalocean.com/?refcode=142755e4323e

The last line of the output tells you how to switch over to using the new host. OSX and Linux users will get an `eval` command to run, while Windows users will get an appropriate command for Powershell or CMD. Copy and paste that command and the current terminal will now be pointing to the remote machine.

From this point forward, all the Docker commands you run will be run against the remote machine. This is seemless as evenm your local install of Docker is handled over an HTTP API, so at worst some of the more complex commands like `docker build` will take a bit longer to run as it packages and transfers files across the internet. Just about everything works fine when talking to a remote machine except host mounted volumes. When you specify a host volume mount, it attempts to mount it from the machine Docker is running on, not your actual local computer.

Once the machine is up and running you can do many different things with it aside from just running Docker commands. We can manipulate the power of the machine to start, stop, and restart it through the driver's individual APIs. If something breaks, you can run `docker-machine provision` to reinstall Docker on a machine. You can copy files to the remote machine with `docker-machine scp`, and if you need to get CLI access directly on the box you can with `docker-machine ssh`.

One big important thing you can do is upgrade the version of Docker that is installed with `docker-machine upgrade`. This is really nice for programmatically handling machine maintance.

Sharing Machine Credentials

The one major downside to Docker Machine is that the configuration is not technically transferable and sharable between team members. All the configuration is generally stored in ~/.docker/machine, so you will need to manually copy that to other people's machines to give them access, and edit the configuration to change all the paths. I would recommend running Docker Machine from a shared server so that everyone can access the servers for things like deployments.

Chapter

8

Swarm Mode

Docker, as we have been using it, has all been running on a single machine. One of the big selling points of Docker is that you can make it much easier to deploy your code, and in this day and age we are not always deploying everything to one machine. There are also the issues of appropriately using resources on hardware, and as hardware gets bigger and bigger we don't want to waste CPU or memory that is just idle.

With Docker Engine 1.12, Docker transitioned a seperate product, Docker Swarm, into an integrated mode called Swarm Mode. Swarm Mode allows us to link together Docker installations into one giant mega-installation, across many different computers, be it physical or virtual. We can start to balance our deployments automatically, or do things like putting together scaling across multiple machines, using our Docker containers. Swarm handles all of the bookkeeping and distribution of containers to the various Docker hosts.

Under Swarm Mode, we run containers against the swarm instead of a specified machine. Swarm Mode will handle finding an appropriate host, starting the image, and any other things it needs to do to get the image running. From there you can query the swarm, or the individual container, to get information such as networking or which host it physically resides on.

Creating a Swarm

The easiest way to get started with Swarm Mode is to create a few test machines with Docker Machine. For this chapter we will create three machines - d4d-manager1, which will control the swarm, and d4d-node1 and d4d-node2, which are just additional nodes to run containers on. Create them of any size that you feel comfortable paying for or running on your setups.

Docker Machine Swarm Options

DO NOT *use the Swarm parameters that Docker Machine provides. It will set up an older, less useful version of Swarm. The current way to install Swarm is via the method we describe here.*

Swarm Mode requires at least one manager, which keeps track of everything, and any number of additional workers. Keep in mind that generally the manager node will also be running containers. To get started, we will need to go onto our d4d-manager1 node and initialize the swarm. This will convert the Docker Engine on this machine over to Swarm Mode. It will also give us a command to run on each of the workers to attach them to the swarm.

```
1.  $ docker-machine ssh manager1
2.  d4d-manager1$ docker swarm init
3.  Swarm initialized: current node (wtcyiri2t8xbeezjlklik8omw) is now a manager.
4.
5.  To add a worker to this swarm, run the following command:
6.
7.      docker swarm join \
8.      --token SWMTKN-1-0u6f9a5wi7n9gseo5kbdu7zegi1n6bogdljc0gtl98oiiul-
zig-8ozfdtwsewpmjnp3a34xal0w1 \
9.      172.16.0.204:2377
10.
11.  To add a manager to this swarm, run 'docker swarm join-token manager' and follow
the instructions.
12.  d4d-manager1$ exit
```

Now we just need to SSH into each of the nodes and join them to the cluster:

```
1. $ docker-machine ssh d4d-node1
2. d4d-node1$ docker swarm join --token [token] 172.16.0.204:2377
3. This node joined a swarm as a worker.
4. d4d-node1$ exit
5. $ docker-machine ssh d4d-node2
6. d4d-node2$ docker swarm join --token [token] 172.16.0.204:2377
7. This node joined a swarm as a worker.
8. d4d-node2$ exit
9. $ docker-machine ls
10. NAME          ACTIVE  DRIVER  STATE    URL                        SWARM   DOCKER   ERRORS
11. d4d-manager1  -       hyperv  Running  tcp://172.16.0.204:2376            v1.13.1
12. d4d-node1     -       hyperv  Running  tcp://172.16.0.201:2376            v1.13.1
13. d4d-node2     -       hyperv  Running  tcp://172.16.0.202:2376            v1.13.1
```

We now have a three node swarm, as easy as that.

Deploying to Swarm Mode

Starting with Docker Engine 1.12, Swarm Mode has become a very attractive alternative to more complicated tools like Kubernetes, and is easing closer to the user friendliness of tools like Rancher. There are a few things we need to be aware of when working with this deployment service.

The first is that, generally, you will be deploying one application to a swarm, and the idea is that the application will scale up and down inside the cluster. You will not be able to deploy two applications that use the same ports. For example, if you are building our demo app that uses port 80 and port 443 in production, you can't turn around and also deploy a Wordpress app that also uses port 80 and 443. Ports are unique per "stack," or collection of images.

The second is that volumes must be configured to be available across the entire swarm, or at least the nodes that you plan on letting containers scale around on. In general, each node must be configured separately to talk to whatever distributed volume system you want to use, be it Flocker, NFS, or something else. This sample project is not taking that into account, so our MySQL container will not scale like other containers would (though, in general, you would not contain your database).

The third is that we cannot build images on demand like we can with a local setup of Docker Compose. This is due to the way that the commands are sent out to the individual nodes and how they handle scaling. They need to be able to just pull an image directly from a repository. For our sample application, we'll need to build our images, push them to a private repo (<hub.docker.com>, in this case), and have our compose file point to them.

And finally, which can be a big turnoff for some companies, is that it is currently all CLI driven. If you need pretty graphs or GUI tools, there are not any available at the moment. You may want to couple a Swarm with a tool like Rancher which can provide those and use Rancher as the visualization of your environment, and manage the actual deployments with Swarm Mode.

There are two different types of deployments you can do with Swarm. The first was introduced in Docker Engine 1.12, which is the "Services" deployment. You manually create networks and define a series of services that will run on the cluster.

The second way is to use it in conjunction with a Docker Compose configuration. This will translate the Docker Compose configuration into a series of services and set everything up for you.

Building The Images

As I mentioned, when deploying to Swarm we cannot build our images on demand like we normally can with Docker Compose. You will need to build your images before deployment and push them to some repository. You can run a private repository that is accessible only to your network, and I would highly recommend that in a production environment.

Docker Hub does have a paid version that uses the same syntax as the second option, and I'll be using the free private repository that they give each user. We'll create a single repository and host two versions of images, one for the ngnix container and the other for the PHP container. This is kind of a roundabout way of getting multiple images up for free to play around with, but I would encourage you to either deploy a free private registry or pay for Docker Hub, which is very inexpensive.

For the purposes of this article, I've set up a private repo on Docker Hub called `dragonmantank/pa-sampleapp`.

There is not anything special to our images, other than that they have to exist as built artifacts before we deploy to Production. If you take a look at our `docker-compose.yml` file, you'll see that the `phpserver` and the `nginx` services both have a `build:` key. This build key is further expanded into a `dockerfile:` and a `context:` setting. We can reproduce this with the `docker build` with the `-f` parameter to pass it a specific Dockerfile to use, and specifying a build context of `./` instead of the path that the Dockerfile lives.

Before we build the images, I'm going to take a quick detour on the subject of tagging images. There are three types of tags that an image can have:

- **word** - php - An officially released image from Docker
- **username/word** - dragonmantank/php - A community image for PHP released by the user dragonmantank on hub.docker.com
- **hostname:port/username/word** - myregistry.lan:5000/dragonmantank/php - An image being pushed to a private registry

You should be very familiar with the first two syntaxes, as your images you build from in your Dockerfiles are probably made from an official or a community release of an image. In a real life production setup you will most likely want to take a look at setting up a private registry since you probably do not want your custom images available to everyone. That is a bit out of scope for this article, but you can take a look at the official documentation.[1]

Of course, after each of the syntaxes you can specify a :version to set up different versions of the same image. For the official PHP image there are options such as php:7-fpm, php:apache, php:5.6-cli, and others that are specific releases of PHP in different setups.

With that out of the way, let's create and tag two images.

```
$ docker build -t dragonmantank/pa-sampleapp:nginx -f docker/nginx/Dockerfile ./
$ docker build -t dragonmantank/pa-sampleapp:phpserver -f docker/php/Dockerfile ./
```

These images should build just like they did under docker-compose, just with a different naming scheme. Next I will log into hub.docker.com through the CLI and push both of them up.

```
$ docker login
$ docker push dragonmantank/pa-sampleapp:phpserver
$ docker push dragonmantank/pa-sampleapp:nginx
```

After a few minutes (or longer, depending on your upload speed), the images will be pushed up to Docker Hub and will be available to any set of machines that I am logged in to.

Deploy with Docker Compose

With Docker Engine 1.13 we now have a docker stack deploy command which can work with Docker Compose files directly without have to create these DAB files or manually manipulate services.

Creating a Production Config

Now that the images are available for pulling down, we can turn our attention to the Docker Compose configuration for Production.

If you look at the files pulled down in git, there are three configuration files for our application. The default one, docker-compose.yml, contains the bare minimum we need to sketch out our application. It has our volume for MySQL, as well as our three defined services (phpserver, nginx, and mysqlserver). We are not specifying any ports or volume mounts as we will let the secondary configurations take care of that.

docker-compose.dev.yml, as we used above, sets up our environment to work on localhost:8080, has our build instructions for the images, and does the host volume mmounting so we can work with the files live. The docker-compose.prod.yml file specifies the exact images that we will use, as well as a more Production-like port forwarding setup on port 80.

[1]   the official documentation.: https://docs.docker.com/registry/deploying/

With all that in mind we now turn to our swarm cluster. Make sure that you are connected to it, and run the `docker login` command against it to log into Docker Hub. As part of the deploy we will tell the swarm master to pass the login information to each of the nodes so that they can all pull from our custom images.

Once you are logged in, go ahead and deploy:

```
$ docker stack deploy -c docker-compose.yml -c docker-compose.prod.yml --with-registry-auth pa-sampleapp
```

If it all goes well, you should end back up at your CLI prompt with no errors. What we told Docker to do was deploy our compose file as a Service Stack, so it coverted our compose files into a stack definition, and deployed it all to a stack named "pa-sampleapp". It also passed along our Docker Hub credentials with the `--with-registry-auth` parameter so that each node could pull down the private images.

We can check the status of the stack by running `docker stack ps pa-sampleapp`:

```
$ docker stack ps pa-sampleapp
ID            NAME                         IMAGE                                 NODE      DESIRED STATE CURRENT STATE
ERROR
tru6e7ttu964  pa-sampleapp_mysqlserver.1   mysql:latest                          node2     Running       Preparing 3 minutes ago
ptk1hnd92l91  pa-sampleapp_phpserver.1     dragonmantank/pa-sampleapp:phpserver  manager1  Running       Preparing 5 minutes ago
ruubz6inerie  pa-sampleapp_nginx.1         dragonmantank/pa-sampleapp:nginx      node1     Running       Preparing 5 minutes ago
```

The deploy can take a few minutes, depending on your internet connection and how many layers there are. We can see though that the swarm deployed our MySQL server to node2, our PHP FPM instance to manager1, and our nginx proxy to node1. Once the images are all out of the "Preparing" state we can actually go to the IP of any of those machines on port 80 and we should be greeted with the web application. In my case, I can go to `http://172.16.0.204`, `http://172.16.0.201`, or `http://172.16.0.202` and all of the traffic is routed to the nginx container, no matter where it needs to go.

Chapter

Dealing with Logs

When your application breaks, what is one thing you need?

Logs.

Using Docker raises a few potential issues when it comes to dealing with logs. Normally there is a file on the server that has the log you are looking for. For PHP this is sometimes Apache's logs in `/var/logs/httpd/` or `/var/www/apache2`. PHP-FPM may have it's own file. You can open that file, tail it, or interact with it in some way to get data from it.

You may also be using remote logging in Production, and sending your log files off somewhere else. As we move to a more "serverless" setup, or at least a more scalable arrangement where servers come and go as they are needed, physical files are not important. Their long-term storage is.

How do we deal with logs when it comes to Docker?

Viewing Container Logs

Branch for this section: `git checkout logs/viewing-logs`

Docker has a built in system for viewing and storing logs. By default, anything that is written to the stdout and stderr streams of a container are stored in a log. This is a JSON file on disk. The trick is getting access to them and working with the logs.

The easiest way to do this is to use the docker logs command. This allows you to view the logs files for an existing container. Run the following command, and after a few minutes stop it with CTRL+C:

```
$ docker run -t --name "testlogs" \
    -v $(pwd):/var/www/html \
    php:alpine php /var/www/html/index.php
```

The output of this command is being written to the stdout and stderr streams of the container. PHP's CLI by default attaches to these streams. In the case of PHP-FPM or mod_php these are redirected to files.

Since the text was streamed to the appropriate places, the output was captured in the Docker logs. We can review the output by running docker logs testlogs.

```
$ docker logs testlogs
Starting an endless counter
Counter: 0
Errors: 302

Counter: 1
Errors: 983
```

docker logs will dump the entire contents of the log file to your screen. On a long running container this could be quite a bit of data. You can view the end of the log file by adding -t X, where X is the number of lines to look back on.

You can also follow a log file as it is written to with the -f flag. This will keep the log file open in your terminal and output any new lines right away. This is very useful if you are watching for errors.

Application Logs

Branch for this section: git checkout logs/php-stdout

Before we dig into working with the logs, we need to think about our application logs. A core idea behind Docker is that containers can be created and destroyed at will. While the container is running we can always hop in and look at logs, but that will not help us long term. We need to move our application's logs from a physical file on disk.

Docker recommends that you log everything to `stdout` and `stderr`. The default logging for Docker takes and uses those two system streams to deal with its own logs, so we will need our application to do that as well. How we go about that depends on our actual images.

If you use the Apache build of the PHP images, it is already set up. Let's start a quick server and take a look at the logs:

```
$ docker run -d --name "testapache" \
    -p 8080:80 \
    -v $(pwd):/var/www/html \
    php:apache
```

You should be able to go to `http://localhost:8080` and see a white page that says "You found me."

The `index.php` file has a call to `error_log()`, which is a function that intentionally logs to whatever the error log is set to. Under Apache, PHP will use whatever Apache is configured to for logging. The Apache config is set to log to a `$APACHE_LOG_DIR/error.log` and `$APACHE_LOG_DIR/access.log`.

The Dockerfile[1] takes this a step further and sets up a softlink for `/dev/stdout` and `stderr` for those logs files:

```
# logs should go to stdout / stderr
RUN set -ex \
    && . "$APACHE_ENVVARS" \
    && ln -sfT /dev/stderr "$APACHE_LOG_DIR/error.log" \
    && ln -sfT /dev/stdout "$APACHE_LOG_DIR/access.log" \
    && ln -sfT /dev/stdout "$APACHE_LOG_DIR/other_vhosts_access.log"
```

Now anything logged to those files will be pushed to `stdout` and `stderr`, just like Docker wants.

We can view the logs of an individual container by running `docker logs`. This will dump all of the logs to the screen. We can make this shorter by using `--tail N`, where N is the number of lines to show from the end. We can also follow the logs with `-f`, and watch as logs are generated.

Let's start to watch the logs:

```
$ docker logs --tail 5 -f testapache
```

and then refresh the website a few times. You should see the error log being generated, as well as the access log showing the page was accessed.

---

[1] Dockerfile: http://phpa.me/docker-for-devs-apache-log

PHP when run from the command line automatically attaches to `stdout` and `stderr`, so there is nothing needed for command line runs.

The PHP-FPM container does something similiar, by setting the `error_log` PHP INI setting to `/proc/self/fd/2`, which on most systems is linked to `/dev/stderr`. `stdout` is not redirected as `stdout` is sent back to the browser.

If you build your own images from scratch, you will need to take this into account.

Logging Drivers and Remote Logging

Branch for this section: `git checkout logs/efk-logging`

So far we've been looking at working with a single container's logs. If you need to dig through them for errors you'll need the help of some other tools. Unix tools like `grep` are great for this sort of work, but it still requires you to have access to the logs.

In a Production environment, you may not have direct access to a container. In fact, the error may have happened on a container that no longer exists. We'll need to think about logging to more than a local JSON file.

Docker allows you to redirect the output of a single container, or the entire system, to something other than the default JSON logger. Docker supports a variety of logging drivers including syslog, journald, gelf, fluentd, and splunk.

Many people are familiar with the ELK stack, which consists of Elasticsearch, Logstash, and Kibana. Docker does not support Logstash, but it does support fluentd, which is a very similiar product.

We can easily redirect output using Docker Compose or `docker run` to a different source than the host machine's logging driver. Since we are using Docker, we can also easily test the EFK stack.

Check out the `logs/ekf-logging` branch of the sample github project. This branch contains a `docker-compose.yml` file and a `Dockerfile` that we can use to transfer all of the output of an Apache installation to Elasticsearch and make it viewable via Kibana. The end of this chapter will also contain these files.

The `docker-compose.yml` file has a new section for the `webserver` service: `logging`.

This key allows you to specify logging options for an individual container. We are going to use this to change the driver to `fluentd`, and specify the location of that server. Each driver has its own set of options, but this will get us working with fluentd. Any log output will no longer be availabel to `docker logs`, as that command relies on the JSON driver.

We create a fluentd service. This will accept all of the logs from a container, and do something with them. We will push in a config file that forwards all the logs to Elasticsearch, which is a document database. This is doing the role that normally Logstash would.

Elasticsearch does not have a GUI, but Kibana is an excellent one to use. It is a web application that works with Elaticsearch and allows for searching of data.

Fire up the stack with `docker-compose up -d`.

For a sanity check, run a quick `docker-compose ps` and make sure everything is running:

```
Name                            Command                     State
-------------------------------------------------------------------
dockerfordevs_elasticsearch_1   /docker-entrypoint.sh elas ...   Up
dockerfordevs_fluentd_1         /bin/sh -c exec fluentd -c ...   Up
dockerfordevs_kibana_1          /docker-entrypoint.sh kibana     Up
dockerfordevs_webserver_1       docker-php-entrypoint apac ...   Up
```

After a few moment, access the webserver at `http://localhost:8080`. Do this a few times to push some logs through fluentd and into Elasticsearch.

Hopefully everything is working. To check, head over to `http://localhost:5601` to access Kibana. It should look like **Figure 9-1**.

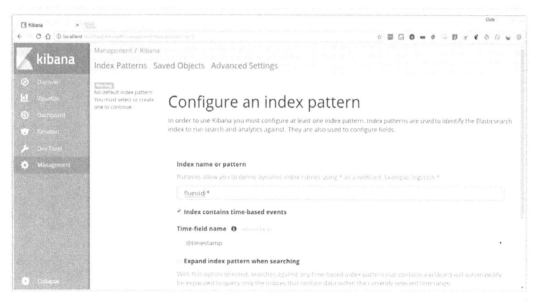

Figure 9-1

Change the "Index name or pattern" to fluentd-*, and the "Time-field name" should auto-matically change to "@timestamp". Scroll down and click 'Create' to set up the index.

If it does not that indicates that something is not working. Check the logs of the fluentd server with docker logs and see if something is not working with fluentd.

Click on the 'Discover' link on the left-hand side. This will bring you to the main search screen, like **Figure 9-2**.

Figure 9-2

You can now search by tags. If you want to search for logs from a specific container, you may try something like container_name: *web*.

From here you can start to explore different logging searches. If you have an error that you are looking for, you can do search for log: [text], where [text] is part of the error. Our index.php script should be writing the string "And an error happens" on every page view, so try searching for something along those lines.

Kibana is a very powerful visualization tool as well. I highly recommend digging into it not only for logging, but visualizing error rates and traffic in Production.

This should get you started with remote logging and it's possibilities. In Production I recommend this sort of setup, though you may want to go with a hosted solution like Logz.io for the actual Elasticsearch and Kibana install. At the very least you can use this as a template for adding EFK logging to your development environment.

Sample EFK files

docker-compose.yml

```
 1.  version: '2'
 2.
 3.  services:
 4.    webserver:
 5.      image: php:apache
 6.      volumes:
 7.        - ./:/var/www/html
 8.      depends_on:
 9.        - fluentd
10.      ports:
11.        - 8080:80
12.      logging:
13.        driver: "fluentd"
14.        options:
15.          fluentd-address: 127.0.0.1:24224
16.          tag: apache.access
17.
18.    fluentd:
19.      build: docker/fluentd
20.      volumes:
21.        - ./docker/fluentd/:/fluentd/etc
22.      ports:
23.        - 24224:24224
24.        - 24224:24224/udp
25.
26.    elasticsearch:
27.      image: elasticsearch
28.      expose:
29.        - 9200
30.      ports:
31.        - 9200:9200
32.
33.    kibana:
34.      image: kibana
35.      ports:
36.        - 5601:5601
```

docker/fluentd/Dockerfile

```
FROM fluent/fluentd

RUN ["gem", "install", "fluent-plugin-elasticsearch", "--no-rdoc", "--no-ri", "--version", "1.9.2"]
```

docker/fluentd/fluent.conf

```
1.  <source>
2.      @type forward
3.      port 24224
4.      bind 0.0.0.0
5.  </source>
6.
7.  <match *.**>
8.      @type copy
9.      <store>
10.         @type elasticsearch
11.         host elasticsearch
12.         port 9200
13.         logstash_format true
14.         logstash_prefix fluentd
15.         logstash_dateformat %Y%m%d
16.         include_tag_key true
17.         type_name access_log
18.         tag_key @log_name
19.         flush_interval 1s
20.     </store>
21.     <store>
22.         @type stdout
23.     </store>
24. </match>
```

Chapter

10

Twelve Factor Applications and Containers

While not specific to Docker itself, the "Twelve Factor Applications" workflow by *https://12factor.net/* should be one of the first things a developer reads and understands.

"Twelve Factor Applications" lays out a set of rules and design decisions that help scalable, cloud enabled applications be easily developed and deployed. Docker also helps scale applications and run them in the cloud, so they go hand-in-hand.

Following the rules laid out in the "Twelve Factor Applications" website will help make sure that your application scales well inside of Docker. There are benefits to using Docker even without taking these into account, but they will certainly help.

Codebase

Code should be stored, and tracked, in a version control system.

It does not matter what version control you use. I like git. Use whatever you want. Just use something.

A good code repository should track who changed what when, allow you to remove those changes, and allow you to mark specific instances of code. If it does not do all of these things, find a new code repository.

Each application should have it's own codebase. Your application may have many moving parts, like a web API, some workers, and some command line utilities, but they all make up a single application. Do not store your CLI tools in one repository, and your API in another. Trust me, it only leads to code being mismanaged.

It may be easier to handle each chunk of the application as its own application. If you do that, treat each resource as a fully separate application. Do not share code between applications, and you will need to start tracking each application as it's own Twelve Factor app.

If you have multiple applications in a repository, it is not following the "Twelve Factor Application" recommendations. I should not have my deployment tools tracked in the same repository as my application.

Track not only your code, but things like Docker Compose and Docker Build files. These are part of your application.

The bottom line is that if you are not using source control, start.

Dependencies

Every application should have their dependencies clearly declared isolated. For modern PHP applications that use Composer this is already taken care for us. Composer creates a `composer.json` file that declares what our dependencies are, and a corresponding `composer.lock` file that declares what versions are currently being used. Each installation of the application gets its own install of the dependencies, isolated from the rest of the system.

For PHP developers this means making sure that you check in both the `composer.json` file and the `composer.lock` file, and installing your dependencies with `composer install` instead of `composer update`.

Checking in the `composer.lock` file ensures that each build of the application will be exactly the same. If we do not check it in, we may end up with slight version differences between builds.

For example, if I develop a branch using version 1.3.3 of a library, and by the time we build it for Production a new version, 1.3.4 is released that fixes a bug I rely on, we have a dependency mismatch. Unless I've pinned `composer.json` to a full specific version (which itself is a bad idea), Composer will always grab the latest version that it can. I may not want that version.

Checking in `composer.lock` ensures that the same versions will be installed, regardless of what the currently available version is. If the lock file has 1.3.3, Composer will install 1.3.3 even if there are newer versions.

If you have a `composer.lock` file then that means you should always use `composer install` to install dependencies. `composer update` will update the lock file, resulting in a situation where each install has potentially different versions.

Docker will help handle our extensions and our "system" dependencies. Creating custom images will ensure that builds are repeatable.

I would suggest tagging image builds so that you can store multiple versions and why the base images are updated. The nature of Docker images means you are not rebuilding them unless necessary, and this helps lock down the knowledge of what dependencies are needed and being used.

Configuration

Anything that can vary between deployments is considered Configuration, and should be part of the environment.

This means that anything like MySQL connection information, or authentication tokens to third party services, needs to be bundled differently than "application" configuration like routes.

"Twelve Factor Applications" recommends storing this data as an environment variable. This way your application can just ask the environment for this information instead of it being bundled with the code. Your staging environment will have a different MySQL host and user than your Production environment, so it makes sense to store that in the environment.

Your PHP application can use something like `getenv()` to read these environment variables. Your application can then move from environment to environment without needing a specific configuration file.

Docker allows you to pass environment variables to containers using the `-e` flag when using the `run` command. Docker Compose allows you to specify an "environment" key for a service. Swarm allow you to also specify environment variables using `-e` when creating a new service.

This is a bit of a break from a traditional PHP application where a configuration file is used. Instead of your release process adding a config file, it will now start a container with the needed environment variables.

Backing Services

Treat all resources as third party resources that will change. Never assume that a service that your application is talking to is a local service, or that it will exist forever.

Docker somewhat enforces this. By using containers the only communication that can happen is network communication. On top of that, since your application can scale you are most likely using hostnames for the communication instead of bare IPs, as you will never know prior to deployment what IP address a container will get.

Your application should be able to handle service outages and changes in a robust manner, as much as can be expected. If you are using a caching layer, make sure your application can handle that caching layer going down. Make sure that your application can handle slow responses from that cache layer. If you can think of a network hiccup, build that into your application.

You want your application to be able to be swtiched from "local" container-backed services to remote third party ones without code changes. Config changes are fine, but your code should never need to be touched just because your MySQL server died and the admin needed to spin up a new one. You should just need to change your configuration.

Take your database, for example. In development you are probably running it in a container, but in Production it is probably sitting in a remote datacenter like Amazon RDS. Your application itself should not care. The only difference should be a config setting that can easily be changed without affecting the code.

If it is a piece of data that can change, then it needs to go into config.

Build, Release, Run

Your application should, and probably already does, go through three distinct phases - Build, Release, and Run. These should be distinct, separate processes. Results from each stage are not modified in any way.

The Build phase should take your code from your code repository, install any needed dependencies, and compile anything that needs compiled. You should be able to do this for any version of your code at any time, and this build should be repeatable and reproduceable. If I build version 4.2 of the application today, when I build version 4.2 in six months from now I should get the same result. The end result of this is normally called a Build Artifact.

For containerized projects, this may mean your build artifact is an image. You can then use this image in the Release phase.

The Release stage takes a build and applies configuration to it. If I want to release version 4.2 to Staging, the release process should take a specific Build Artifact and package it with the needed configuration. This is then "released" whenever it needs to go.

Releases should be tagged in a unique way. This can be with a timestamp, or a release number. This differs from a Version number, which is an indication of something in the codebase. Releases are never modified once they are created. If configuration needs changed, this results in a new release.

As Releases normally just add configuration, a "release" may just be a number and not a physical thing. You may use Environmental Variables for configuration, which makes it so the Build Artifact container is never modfiied with a real configuration file change.

Once the code has been "released" into a new environment, it is in the Run stage. Your code is active and hopefully doing what you intended it to do. Code in the Run stage is never modified. Changes to code need to be made at the repository, a new Build created, and a new Release invoked.

Your build process and release process should be made container aware when possible. You may elect to have Build Artifacts be a simple tar file that has your code, and the Release stage creates an image.

No matter what you do, think of these steps as three different stages, and keep workflows in place to enforce that separation.

Processes

Applications should be thought of as Processes, not as a single entity.

I have stressed that quite a bit throughout the book. You need to break your application down into the processes that are being run, not just "I'm running Wordpress." Running Wordpress invovles a web server and a database server, at the very least.

Each process should be self contained and stateless. My application should run with one, or twenty, PHP-FPM processes. It should run with one, or one hundred, nginx processes. None of the processes need to know about any other specific process, and none of them need to share state.

My nginx process only needs to know that there are PHP-FPM processes available, as it has a hard dependency on PHP. It should not care how many of them there are, and whether or not it has the ability to talk to the same process over and over. As long as a PHP-FPM process is available, that is all that should matter.

PHP should strive as much as possible to be not be tied to a local setup. Sessions should be stored in a globally reachable place like a database, or memcached. User editable files need to be made available to all the processes, so should be stored in such a way that no individual container is needed to access the files. Store them in S3, or GlusterFS, or something else.

Processes are a first class citizen in "Twelve Factor Applications." Docker will handle helping split things into Processes, but you will need to make sure your application handles that as well.

Port Binding

Services should bind to a port, and be accessible over a port. Do not run services over local unix sockets.

For PHP this means running as part of the web server, like mod_php for Apache, or running PHP-FPM on a TCP port.

Docker can expose those ports through the container. You will need to do this anyway to allow containers to talk to each other. Tools like Docker Swarm go a step further and add a routing layer as well, which helps map a public port, like port 80, to an internal port on the containers.

Your containers shouldn't really care what port they are bound to, but should make sure to make themselves available over a port.

Concurrency

"Twelve Factor Applications" recommends scaling should be handled at the process level, not the application level. This means that if we need to handle increased user traffic, we first decide where our bottleneck is, and then scale just that.

For PHP applications this normally means the database. As I do not recommend running databases in containers for Production, scale that through normal means by adding or removing secondary services. Services like Amazon's RDS help with this.

If your bottleneck is PHP itself, scale out the PHP process if you are using PHP-FPM. If you are using something like Swarm for deployment, service discovery will automatically add the new containers into the available pool of processes and start to route traffic to them. If nginx is the bottleneck, do the same thing.

The idea is that you scale up by the individual process. You should not have to deploy an entire additional installation of your application. If you do, your application will not handle concurrency very well.

Docker and Docker Swarm make this easy, but your application itself must be built in a way that additional processes does not break it. Sessions should be handled in the database, or some sort of shared mechanism. User uploaded data like files and images should be in a shared space like a GlusterFS install, or S3. You may need to modify your application to allow for this.

Disposibility

You need to make sure that your application can be scaled up, and down, gracefully and quickly without the application being impacted. By using containers your processes should start and shut down very quickly. If you tried scaling with Composer you can see this right away, and with Swarm the only slowdown is if a new node needs to download an image.

When a container is shut down, Docker sends a `SIGTERM` command. It does not just instantly kill the process. `SIGTERM` simply tells the application that it needs to shut down, and to terminate normally. For web servers, this means cleaning up any last minute deals with connections and shutting down. PHP will run through anything set up through `register_shutdown_function()` to clean up. For most web processes this is probably fine since a web process receives a connection, generates a response, and returns it very quickly.

What does your application do?

There will be a good chance your web application will have some kind of background or long running tasks it needs to deal with. If you are running command line applications in a container, such as a queue worker, you will need ot make sure that they handle receiving a `SIGTERM` in a way that does not break your application. "Signaling PHP", by Cal Evans[1] is a great resource for handling various types of signals command line applications may recieve.

You do not want data corruption or processing workflows to break just because your application does not know how to shut down cleanly.

Development/Production Parity

All your environments, not just Development and Production, should be configured and run the same as much as possible. This reduces the amount of bugs that happen due to environmental differences (even between Linux distributions).

I will give you a real world example. The application I work on day-to-day does not use Docker, for a very good reason. Each developer was allowed to set up their development environment how they wanted. We deploy to CentOS 6 and, at the time, PHP 5.4. We had a developer create a new feature that relied on a CLI call to create a zip file. Worked great on his machine, and the code was pushed up. Made it to testing and it refused to work.

He was using Mac OSX, which uses a different userland than CentOS. This meant a slight discrepency in the arguments that can be passed to the `zip` command, as the BSD userland that OSX uses is different than the GNU userland that CentOS uses. Thankfully we caught it before it went to Production, but because Development and Production were different we ran into issues.

Docker alleviates this. Your developers can, and should, use the same base images that your Production images are built on. This ensures that they are running on the same constraints that Production is.

You can also use Docker Compose `docker-compose.yml` files to create stacks for deployment. This ensures that Compose builds the applications and links the same way across the board.

Do not use or rely on raw `docker run` commands to set anything up. `docker run` is great for one-off testing or prototyping, but make sure that anything that is being used for real, production work is stored in Docker Compose and in the same images that will be used for Production.

"Twelve Factor Applications" also makes mention to use the same services in all environments. Docker makes that easy since you can run all your services on any developer machine, and since everything is stored in containers your developers do not have to install local versions of services.

Logs

Your application should not care about handling logs, or where they are stored.

For PHP we do not usually worry about this, as log handling is more of a function of the web server. Docker suggests logging everything to `stdout` and `stderr`, and your application

---

[1]   "Signaling PHP", by Cal Evans: https://leanpub.com/signalingphp

should do the same. If you are writing a command line utility, it should not log to a file. It should log to stdout and stderr.

How do you read logs then, if they are tied to a specific container's stdout? You will need to send them somewhere.

There are a few ways you can do this. I've outlined one way in the "Docker Logging" chapter, which deals with sending logs to Elastic Search, but there are tools like Filebeat[2] which can grab logs from a host to send to ElasticSearch.

Your application should not care about this in any way however. If should just write to stdout and stderr, and let the host or Docker handle what happens to those logs.

Admin Processes

Administrative processes should run in the same environment as the actual application. If you need to run database migrations, your admin command for that should run in the same setup as the environment you are running against. Run it in a container, in the same pool and network. Use the same base images as the application itself.

One thing you may want to look into is creating images to go along-side your base application images. This way during builds you know that your tools will be packaged with all their dependencies, and are in-sync with the rest of the application.

If you use something like Symfony's command line packages then you can build these tools directly into your application. You can then spin up a container that executes just that single command and shuts down once it completes.

What you want to avoid is needing to have admins and devops need to replicate the "ssh in and run a command on the server" process. There is no SSH, and you cannot even guarantee that the container they used previously even still exists.

Containerize administrative needs, and run them just like you would any other container.

[2]   Filebeat: https://www.elastic.co/products/beats/filebeat

Chapter

11

Command Cheatsheets

Images
- `docker images` - List out all of the images on the host machine
- `docker rmi [image]` - Removes a Docker image if no containers are using it
- `docker pull [image][:tag]` - Downloads a tag from a registry

Containers
- `docker ps [-a]` - Lists containers currently on the system
- `docker run [options] IMAGE [command] [arguments]` - Creates a container and starts it
- `docker start [name]` - Starts a running container
- `docker attach [name]` - Re-connects to a container that is running in the background
- `docker stop [name]` - Stops a running container
- `docker rm [-f] [name]` - Removes a container
- `docker create [options] IMAGE [command] [arguments]` - Creates a container

docker-machine

- `docker-machine create --driver [driver] [machinename]` - Creates a new Docker Machine
- `docker-machine ls` - Lists all of the available machines
- `docker-machine env [machinename]` - Returns the information about a machine
- `docker-machine start [machinename]` - Starts a stopped machine
- `docker-machine stop [machinename` - Stops a running machine
- `docker-machine restart [machinename]` - Restarts a machine
- `docker-machine ip [machinename]` - Returns the IP of a machine
- `docker-machine rm [machinename]` - Destroys a machine
- `docker-machine ssh [machinename]` - SSH's into a machine

docker-compose

- `docker-compose up -d` - Builds a project
- `docker-compose ps` - Shows container information for project
- `docker-compose rm` - Destroys a project
- `docker-compose stop` - Stops a project
- `docker-compose scale [container]=[size]` - Adds or removes containers for scaling

Index

www.ingramcontent.com/pod-product-compliance
Lightning Source LLC
LaVergne TN
LVHW080101070326
832902LV00014B/2366